GEOLOGY UNFOLDED

An Illustrated Guide to the Geology of Utah's National Parks

Thomas H. Morris, Scott M. Ritter, Dallin P. Laycock

GEOLOGY UNFOLDED - An Illustrated Guide to the Geology of Utah's National Parks:
Prepared by Thomas H. Morris (Ph. D.), Scott M. Ritter (Ph. D.), and Dallin P. Laycock; Published at BYU Press under Copyright to Geology Unfolded LLC, 2012. Reorders: Phone (801) 358-5308; General info.: Phone (801) 226-5479 or email: publications@geologyunfolded.com or see us online at: geologyunfolded.com.
Photo of Earth as shown on the cover and this page from www.everystockphoto.com. See creativecommons.org.
ISBN 978-0-8425-2766-8.

TABLE OF CONTENTS:

PREFACE:

The national parks of Utah are particularly "geologic" in nature. They are all located within the Colorado Plateau, an elevated, relatively dry geologic province of the North America continent. As such, vegetative cover is minimal and the sedimentary rocks that dominate the plateau are strikingly exposed. The variety of landscapes observed in the parks result from three broad geologic processes: 1- the deposition of sediment in diverse environments which created the rock strata, 2- the uplift and deformation of the strata by continental-scale tectonic forces, and 3- on-going weathering and erosion of the strata exposed at Earth's surface.

This book was written for the non-geologist and geologist alike who desire a fundamental understanding of each park's geology ... the heart of the landscape. Each chapter succinctly explains the geologic processes that created the landscapes and geologic features in Utah's parks. We hope to answer some of the most frequently asked questions.

This book is the culmination of a number of individual tri-folded brochures that we wrote over the past six years. These brochures explain the geology of each park and of the Glen Canyon National Recreation Area. We titled that series of brochures "GEOLOGY UNFOLDED". Each includes a blend of aesthetically-pleasing aerial and ground photographs designed to illuminate the reader's understanding of the geology. Additionally, we incorporated several figures that illustrate the processes that formed various geologic features. Each brochure also includes a stratigraphic column that serves as a reference for the geologic formations in each park, including their ages. In addition to the materials from the "GEOLOGY UNFOLDED" series, we have included the following in this book:

- Utah Highway Map including the National Parks
- An Overview of the Geologic Time Scale
- A Reader's Guide to Understanding a Stratigraphic Column
- An Illustrated Glossary – to clarify important or technical terms that may be unfamiliar to the non-geologist. We also refer the reader to the park(s) that displays the defined term most prominently.

The idea for the brochures resulted from the senior author's sabbatical to New Zealand in 2002. Several of the national parks there had a similarly styled pamphlet that summarized its geology. These were most satisfying. We hope you find this book similarly satisfying and insightful. Ultimately it is designed to enhance your visit and "Unfold" the geologic story behind Utah's spectacular national parks!

In Memory of Ken Hamblin: 1928-2009

Ken Hamblin was full of life. He was a beloved husband, father, instructor, mentor, colleague, and... geologist. Pictured here in his heaven, within the "Great Unknown" of the Grand Canyon.

Born in Lyman, Wyoming Ken's passion of geology and particularly the creation and evolution of landscapes followed him throughout his life. This passion placed him into the fuselage of many small aircraft where he could be found snapping photographs. This occurred most often during the dawn and dusk when low angle light cast magical shadows and hues that best illustrated the geology of the landscape. Ken had the "artist's touch" of capturing the perfect shot and then being able to illustrate the subsurface geology associated with the landscape. A number of those photographs are incorporated in this book. They were all freely given to us the authors, along with advise and editing.

Ken was an accomplished writer and illustrator of geology. For decades his introductory physical geology textbook "Earth's Dynamic Systems" was the industry standard and student's bible. Its success, in part, stemmed from its many vivid photographs and illustrations. His last two books, "Beyond the Visible Landscape: Aerial Panoramas of Utah's Geology" and "Anatomy of the Grand Canyon: Panoramas of the Canyon's Geology" represent his legacy, passion, and farewell to those who learned from him. One former student represents the feelings of hundreds:

"... He instilled an enthusiasm that remains with me to this day. He had the unique ability to make the geology come alive for the student. One constant in his classes were the stunning, visual images of geology from around the world. They created a lasting impression that, in part, inspired me to pursue a lifelong career in geophysical visualization. I will forever be grateful to the motivation and encouragement he provided .... For Ken, the Grand Canyon was a living, vivid portrait of the earth, and a canvas upon which he illustrated its history. Your legacy is the folks who see and appreciate geology in ways they never imagined. Thanks Ken for the inspiration."

- R. William (Bill) Keach II

Geology has lost a passionate voyager. Humankind has lost "a good man". We will all miss you Ken.

The Authors

UTAH HIGHWAY MAP:

To aid in planning your trip to Utah's national parks, please enjoy this highway map of Utah. This map is not comprehensive, but it includes the major highways and urban centers.

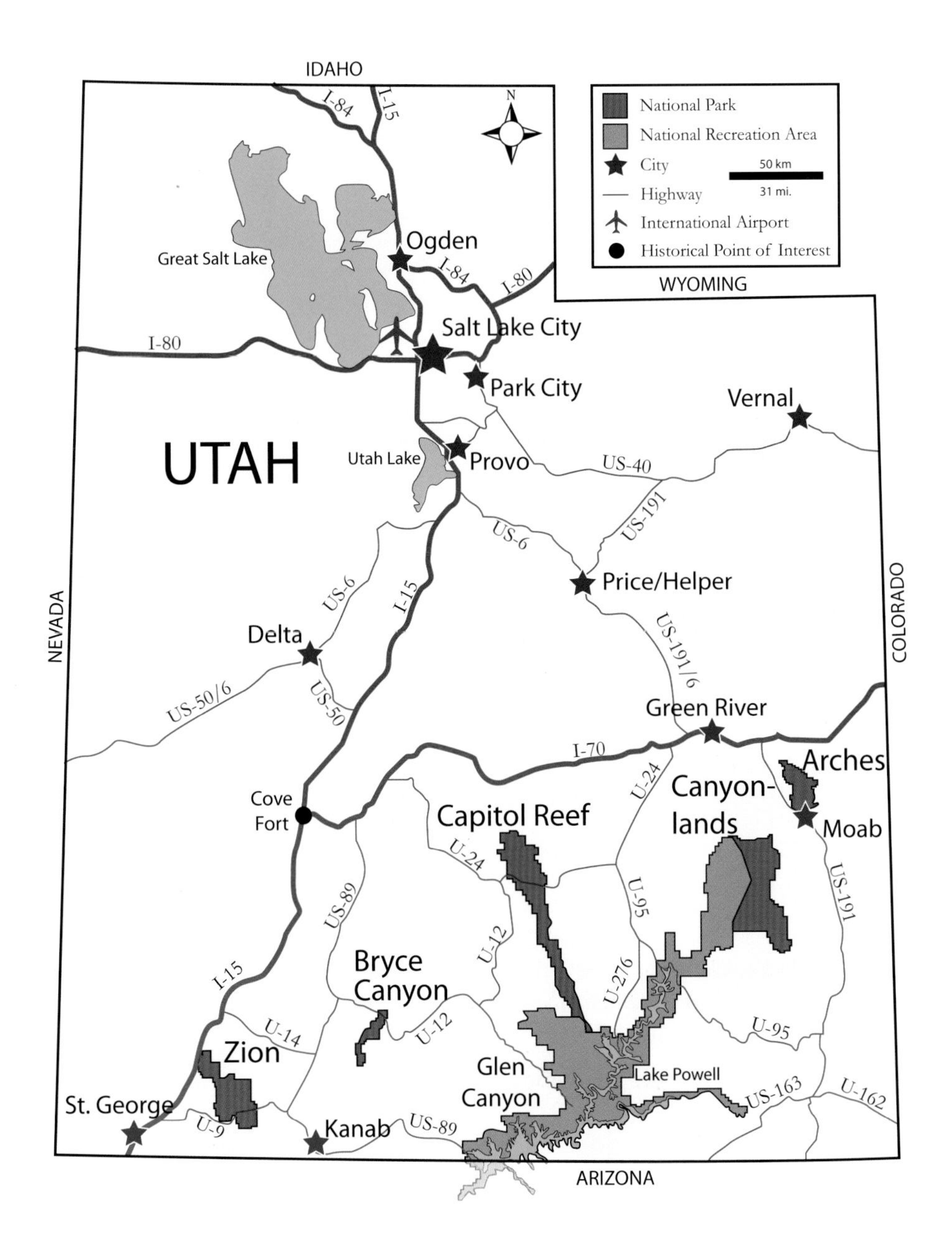

GEOLOGIC TIME SCALE:

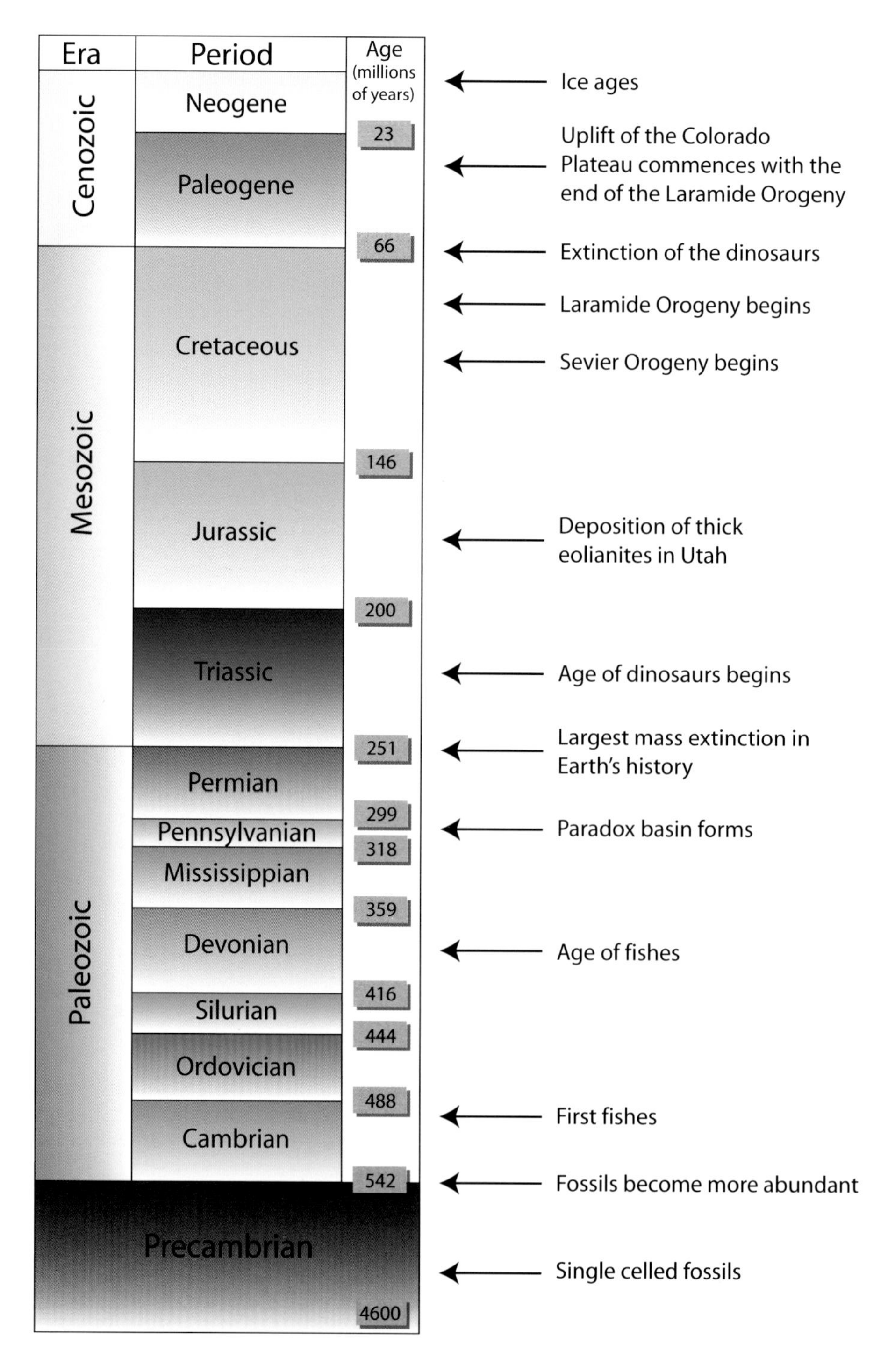

GUIDE TO UNDERSTANDING STRATIGRAPHIC COLUMNS:

Geologists use illustrations known as "Stratigraphic Columns" or "Strat Columns" to show how the rocks of a certain area relate to each other. In this book, they are located at the end of each chapter. Use this key to help you understand what they are saying about the geology of each area.

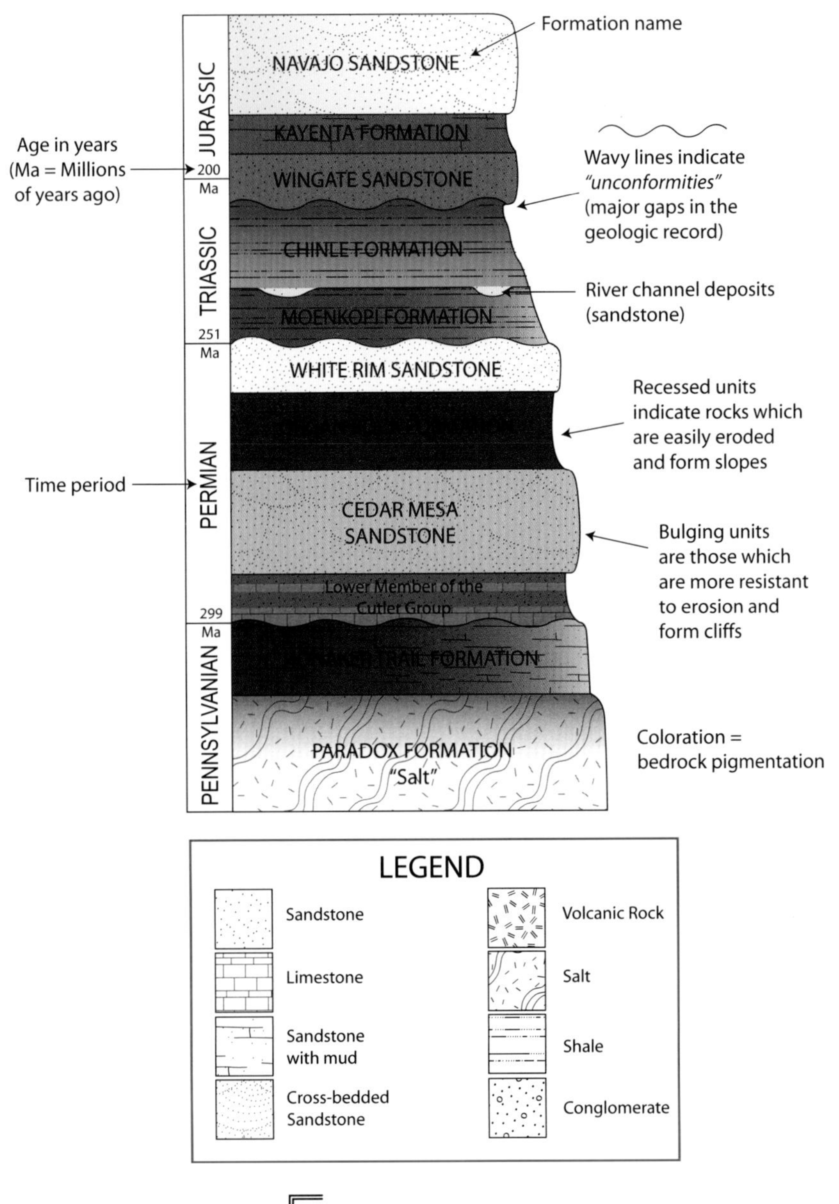

Chapter 1
Arches National Park

Introduction

The red rock desert region of eastern Utah, set aside as Arches National Park in 1971, contains the highest concentration of natural arches in the country, among which is the world famous Delicate Arch. Over 2000 rock spans ranging from three to 306 feet (one to 93 m) in length adorn the 76,359-acre park. These and other scenic rock formations have been carved principally from four Jurassic-aged strata: the Navajo Sandstone, the Dewey Bridge Member of the Carmel Formation, the Slick Rock Member of the Entrada Sandstone and the Moab Member of the Curtis Formation. The unusual park scenery is attributed to the region's unique history, a fortunate and complex interaction between deposition, deformation (folding and fracturing), and erosion that began in the middle of the Pennsylvanian Period (more than 300 Ma) and that continues to shape the landscape today.

Arches at a Glance:

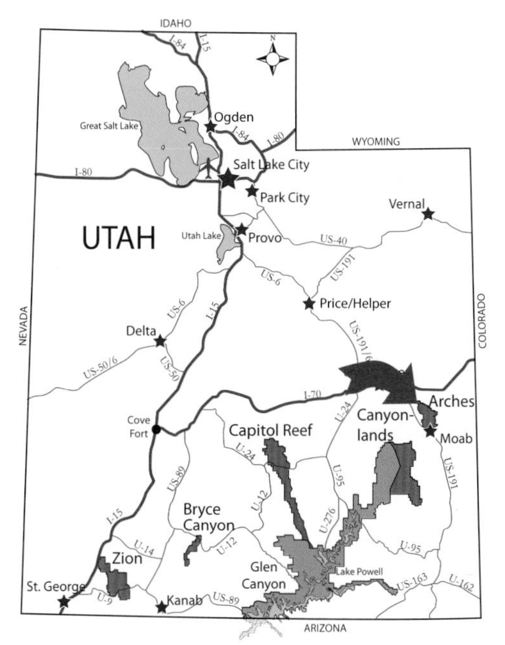

FAST FACTS:

Created as a National Monument:	1929
Given National Park Status:	1971
Climate:	Arid (Desert)
Land Area:	76,359 acres
Geological Province:	Colorado Plateau
Age of Exposed Bedrock:	Pennsylvanian (300 Ma) – Cretaceous (75 Ma)
Arch-forming Strata:	Carmel Formation, Entrada Sandstone, and Curtis Formation
Number of Documented Arches:	>2000
Elevation:	4021 to 5653 feet (1250 to 1760 m) above sea level

What Rocks Form the Scenery?

Four Jurassic formations are responsible for most of the park's magnificent scenery. These are, from bottom to top, the Navajo Sandstone, the Dewey Bridge Member of the Carmel Formation, the Slick Rock Member of the Entrada Sandstone, and the Moab Member of the Curtis Formation ***(Figure 1; see stratigraphic chart on page 15)***. From the Visitor Center to The Windows, the main road runs along the top of the cross-bedded Navajo Sandstone near its contact with the reddish-brown Dewey Bridge Member. Looking east toward the La Sal Mountains from this stretch of road, the view is dominated by domes and canyons carved into the Navajo Sandstone, a remnant of an ancient sandy desert.

The park's main attractions have been carved from **strata** of the San Rafael Group. Crinkly reddish-brown siltstone and mudstone beds of the Dewey Bridge Member of the Carmel Formation form the base of most **pinnacles**, towers, and walls. The Dewey Bridge Member represents tidal flats and tidal channels of an ancient shoreline system. There is a sharp boundary between the Dewey Bridge and the massive coastal dune complex sandstone of the overlying Slick Rock Member of the Entrada Sandstone. Weathering and erosion have exploited this plane of weakness to form alcoves and arches ***(Figure 2)***. In the Devils Garden, arches (e.g. Skyline Arch) developed on horizontal seams within the massive Slick Rock Member itself. The Entrada Sandstone is overlain by light tan, cross-bedded to planar-bedded sandstone comprising the Moab Member of the Curtis Formation. The Moab Member represents marine reworking of the coastal dune complex. The light colored beds of the Moab Member form the top of the upper span of Delicate Arch ***(see photo on page 8)*** and the white caprock of **fins** and walls in the Devils Garden. Strata younger than the Moab Member of the Curtis Formation (Morrison through Mancos) are preserved only in Salt Valley and Cache Valley, located in the northwest and east-central areas of the park, respectively. The road between the Salt Valley Overlook and Delicate Arch passes among steeply dipping strata of these collapsed blocks (see discussion below).

Figure 1 (below): *Photo panorama of the relevant Jurassic stratigraphy of Arches National Park. This photo was taken just outside of the park at the type section of the Dewey Bridge Member of the Carmel Formation.*

Figure 2 (below): *Westward view of "The Windows" representing an intermediate stage of arch development (see Figure 3). The base of these arches is formed at the contact between the Dewey Bridge Member (Carmel) and the overlying Slick Rock Member (Entrada). Both of these stratigraphic formations are part of the San Rafael Group as found on the Colorado Plateau.*

How Do Arches Form?

Figure 3 (right): Illustrations depicting the formation of arches with time. A) Early Stage: Erosion of vertical fractures create fins. B and C) Intermediate Stage: Water perches along horizontal contacts between impermeable rock layers below (Dewey Bridge Member) and more permeable rocks above (Slick Rock Member). Eventually, as cements dissolve and grains erode, a small opening is created through the fin. The window is expanded through freeze/thaw cycles and associated rockfall on the bottom side of the arch. D) Late Stage: Finally the arch collapses leaving the pedestal of the arch as a standing tower or pinnacle.

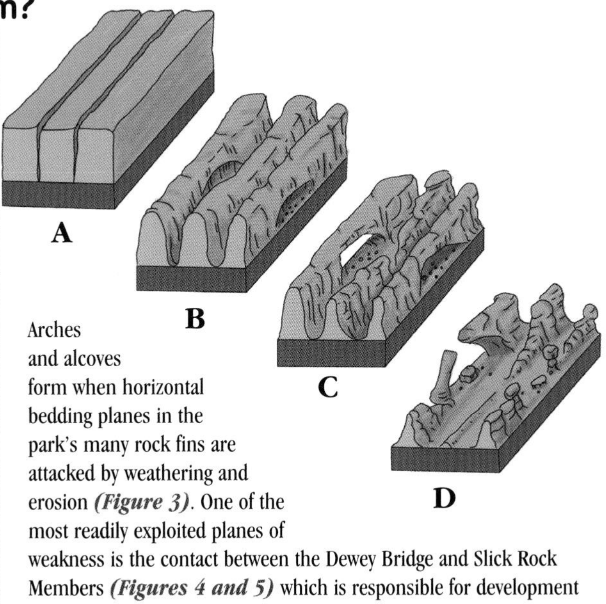

Arches and alcoves form when horizontal bedding planes in the park's many rock fins are attacked by weathering and erosion ***(Figure 3)***. One of the most readily exploited planes of weakness is the contact between the Dewey Bridge and Slick Rock Members ***(Figures 4 and 5)*** which is responsible for development of arches in the Windows region of the park. Slightly acidic groundwater percolates along this plane dissolving calcite **cement** and releasing grains of sand and silt. Over time the fin is breached and a small horizontal opening is created at the contact between the Dewey Bridge and Slick Rock Members ***(Figure 3B)***. Gravity-induced fractures develop in the unsupported sandstone above the breach allowing blocks of Slick Rock sandstone to fall from the bottom of the opening. Continued spalling along these curved fractures causes the opening to grow and to attain its characteristic shape ***(Figure 3C)***. Once in excess of three feet (1 meter) in either the horizontal or vertical dimension, the opening qualifies as an **arch**. Landscape Arch, the longest in the park, has an opening of 306 feet (93 meters).

Figure 4 (right): "Park Avenue" illustrates an early stage of arch development (see Figure 3). Here the rock fins have formed from deep erosion of fracture sets but openings or arches have not yet formed. Note the contact between the wrinkled beds of the Dewey Bridge Member of the Carmel Formation (near the base) and the more massive Slick Rock Member of the Entrada Sandstone (above).

Figure 5 (bottom right): Composed of the Slick Rock Member of the Entrada Sandstone, "Balanced Rock" sits on the contact of the Dewey Bridge Member of the Carmel Formation. It is one of the many pinnacles in the park that illustrates the differential erosion of the two members. Although Balanced Rock may represent a collapsed arch, late stage arch development (see Figure 3) may best be visualized at Sheep Rock in the Courhouse Towers area.

Why Are There So Many Arches Here?

Although the Carmel, Entrada, and Curtis Formations are present throughout much of eastern Utah, only here, north of Moab, are they sculpted into such a remarkable number of fins, pinnacles, alcoves, and arches. The high concentration of closely spaced **fractures** that promotes arch development (see previous section) is explained by uplift and then sagging of brittle sandstones of the Dewey Bridge, Slick Rock and Moab Tongue Members across the tops of the salt-cored Moab and Salt Valley **anticlines**. During the Pennsylvanian Period (320 to 300 million years ago), this area was part of the Paradox basin, an inland sea that dried up intermittently, leaving behind a thick accumulation (5000 feet, 1525 meters) of layered marine salt. Loading by deposition of subsequent Permian through Triassic layers caused the ductile salt to flow to areas of lower pressure—principally along parallel fault lines. Gradually, salt accumulations or **salt walls** up to 10,000 feet (3050 m) thick, three miles (5 km) wide, and 70 miles (110 km) long developed beneath the northwest-trending Moab and Salt Valley faults. The area between sagged in response to withdrawal of the subsurface salt ***(Figure 6)***. Approximately 60 million years ago, and long after the salt walls had been emplaced, compressive forces associated with the ongoing Late Cretaceous-Early Paleogene **Laramide Orogeny** warped the region into anticlinal folds. These fold axes were superimposed over the Moab and Salt Valley salt walls resulting in the **salt-cored anticlines** seen today. Upward bulging of the brittle Dewey Bridge, Slick Rock, and Moab sandstones over the crests of these structures resulted in the formation of closely spaced fracture systems that roughly parallel the axes of the salt anticlines ***(Figure 7)***. Additional widening of fractures occurred in response to dissolution of the upper portions of the salt walls during the past 15 million years ***(Figure 8)***. During uplift of the Colorado Plateau, removal of overlying rocks allowed fresh water to seep into and dissolve the near surface portion of the salt wall. Once undermined, the rocks forming the roof of the salt wall, in this case the Moenkopi through Mancos Formations, collapsed into the void to form a topographic low (or valley) along the crest of the structure. These valleys are then essentially **collapsed salt-cored**

__Figure 6 (below):__ Thick layers of subsurface salt began to flow to zones of lower pressure after burial. Through eons of geologic time (estimated at approximately 75 million years) this vertical salt movement created walls which pushed up overlying rocks layers and eventually served as the core of surface folds (salt-cored anticlines). Modified from Doelling, 2000.

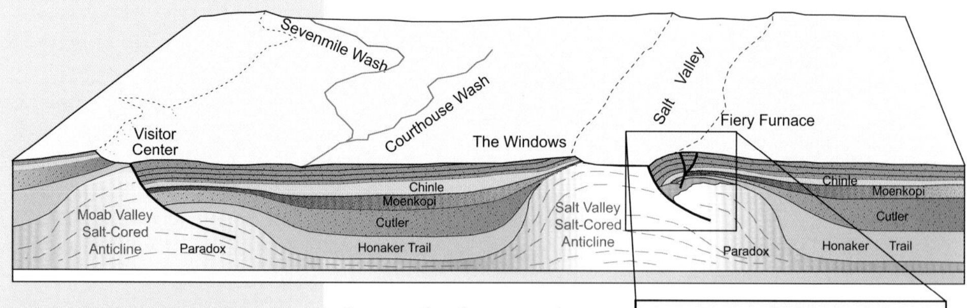

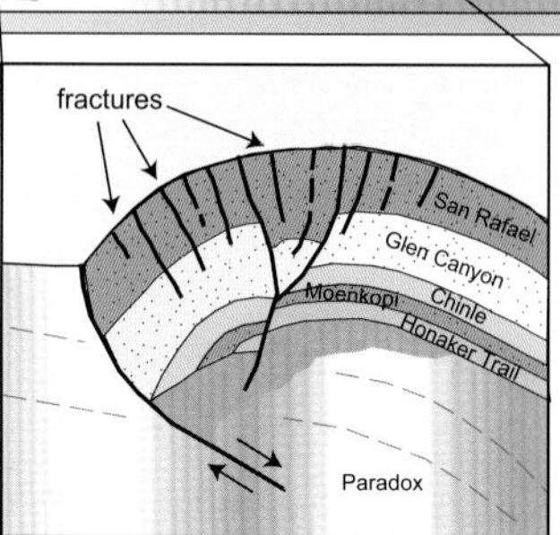

__Figure 7 (inset):__ Along collapsed portions of the salt wall crest the overlying strata (including the Jurassic-aged rocks of Arches N.P.) had to "roll over" into the collapsed zone which created tight folds. This caused the rocks to deform in a brittle manner creating new fractures while widening older ones. Erosion of fractures eventually created "fins" of standing rock that set up the conditions for the creation of arches.

Continued on the next page...

anticlines. Thus, it is not by chance that the Fiery Furnace, Devils Garden, and Klondike Bluffs, those areas with the highest concentrations of arches and fins, are located on the margins of Salt Valley.

Why then are there so many fins and arches at Arches National Park? These features result from the chance combination of soft, ductile Pennsylvanian-aged salt, brittle Jurassic sandstone, and prolonged physical and chemical weathering of these rocks through eons of geological time.

Figure 8 (right): *Southward aerial view of the east flank of the Salt Valley anticline illustrating fracture sets that parallel the rim of the anticline. Deeply eroded fractures in the Slick Rock Member of the Entrada Sandstone (red rocks in center of photo) create the "fins" needed for arch formation. White rocks on left side of photo represent the Moab Member of the Curtis Formation. Right side of photo represents a collapsed salt-cored anticline known as Salt Valley.*

Geological Hazards - Beware of Falling Rock!

Chemical and physical weathering of sandstone loosens blocks of rock along joints and fractures, making them susceptible to the effects of gravity. Blocks both large and small litter the base of cliffs throughout the park, evidence of the ongoing evolution of the park's landscape ***(Figure 9)***. **Rockfalls**, though rarely experienced by visitors, do occur and can pose a hazard to park visitors, especially after heavy rains or on winter afternoons. During the night of August 8, 2008 along the Devil's Garden Trail, Wall Arch collapsed. Formerly it had a 55 foot span and was ranked as the twelveth largest arch in the park.

Hikers occasionally underestimate the pull of gravity on steep-sided slopes, which can create another hazard. Accidents have been known to occur simply by the hiker walking off an ever-increasing rock slope such as a rounded sandstone wall or dome. They reach "the point of no return" wherein gravity overtakes the frictional force holding them to the rock. Loose sand grains can act as miniature ball bearings that reduce the frictional force holding you to the rock. Remember that it is illegal to walk on the arches—for good reason!

Figure 9 (right): *The force of gravity is a major player in the evolution of any landscape. Fracture sets, accelerated erosion, and undercutting of softer rock beneath more resistant rock can create the conditions that create rockfalls. Evidence of rockfalls is common throughout the park.*

How Did the La Sal Mountains Form?

Figure 10 (below): The La Sal Mountains create a scenic backdrop for Arches National Park. The La Sals represent an erosional remnant of ponded magmatic material that was embedded well beneath the surface of Earth approximately 28 Million years ago.

The La Sal Mountains can be seen from many places in Arches National Park ***(Figure 10)***. Their alpine grandeur forms a striking backdrop to the desert scenery. The La Sal Mountains are an **igneous** rock-cored uplift known as a **laccolith**. This dome-shaped igneous rock body intruded (or injected itself) beneath the Earth's surface approximately 28 Ma. The **magma** was injected into the core of a salt anticline located east of Arches National Park. Erosion has removed the sedimentary strata that once surrounded the magma chamber, etching the resistant igneous rocks into the high peaks that reach over 12,000 feet (3660 m) in elevation.

Why Are Layers of the Dewey Bridge So Crinkly?

Figure 11 (below): Enlargement of a portion of Figure 1 illustrating sediment loading of the Slick Rock (Entrada) downward into the Dewey Bridge (Carmel). Sediment loading is one type of soft sediment deformation that is common to the Jurassic rocks of the Dewey Bridge and Slick Rock Members.

This distinctive contorted bedding is due to **soft sediment deformation**. Simply put, these wrinkles formed before the tidal flat mud, silt and sand of the Dewey Bridge had been completely cemented and solidified. Irregularities or wrinkles in the boundary between the Dewey Bridge and Slick Rock suggest that sediment loading (deposition of the sand that would eventually be cemented into the Slick Rock sandstone) may have caused the underlying mud and silt to flow ***(Figure 11)***.

Alternatively, soft sediment deformation may be triggered by seismic events. Seismic activity at this time in Earth history could have resulted from earthquakes associated with oceanic/continental **tectonic** movement or possibly by meteor impacts in the area.

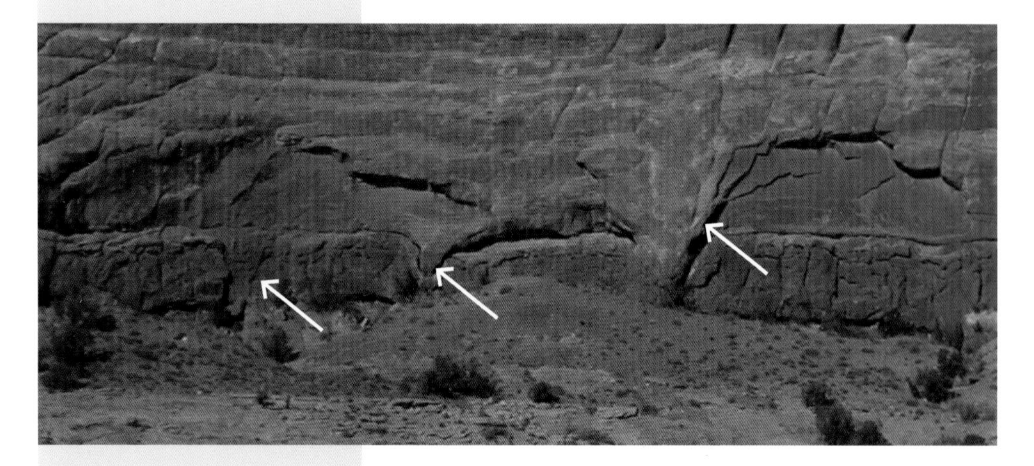

Bedrock Strata of Arches National Park

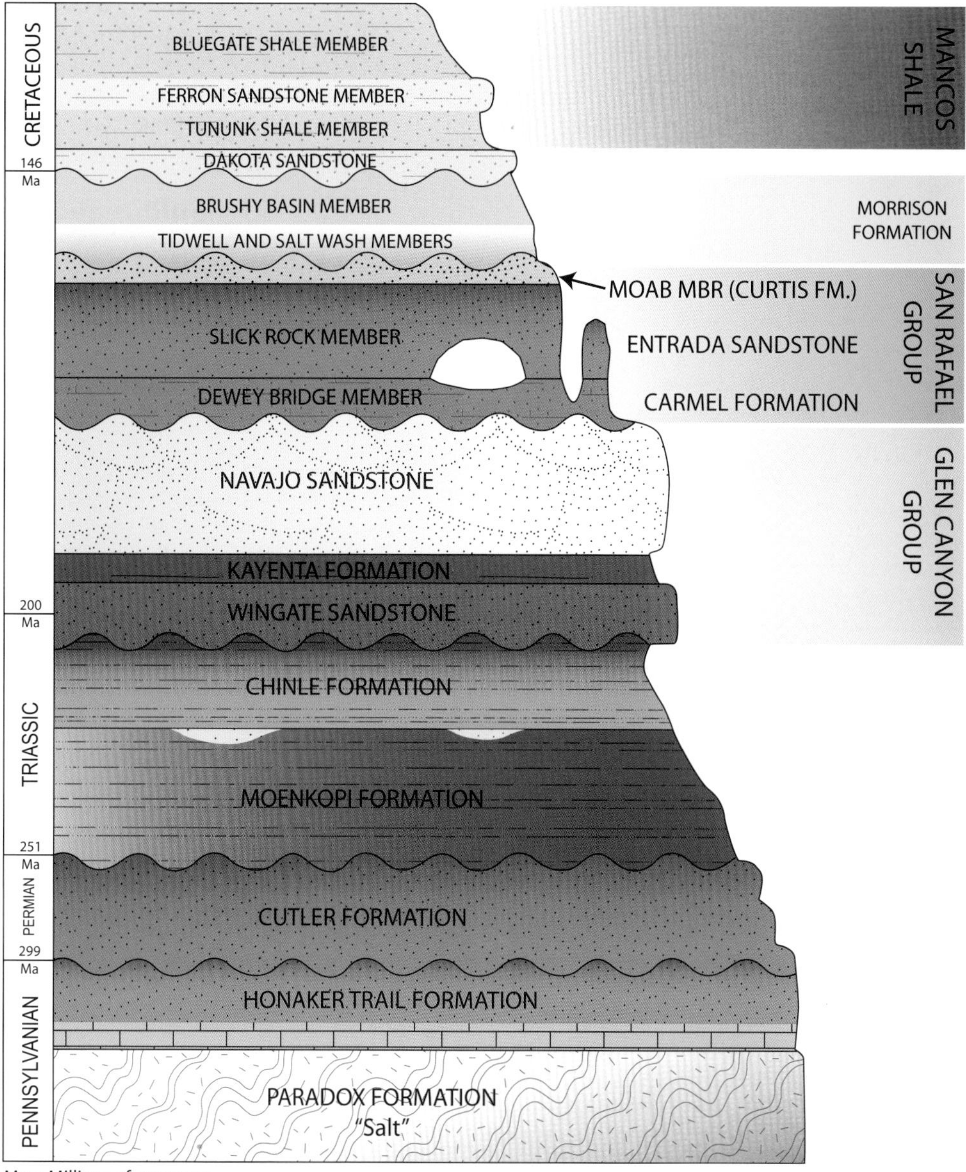

Ma = Millions of years ago

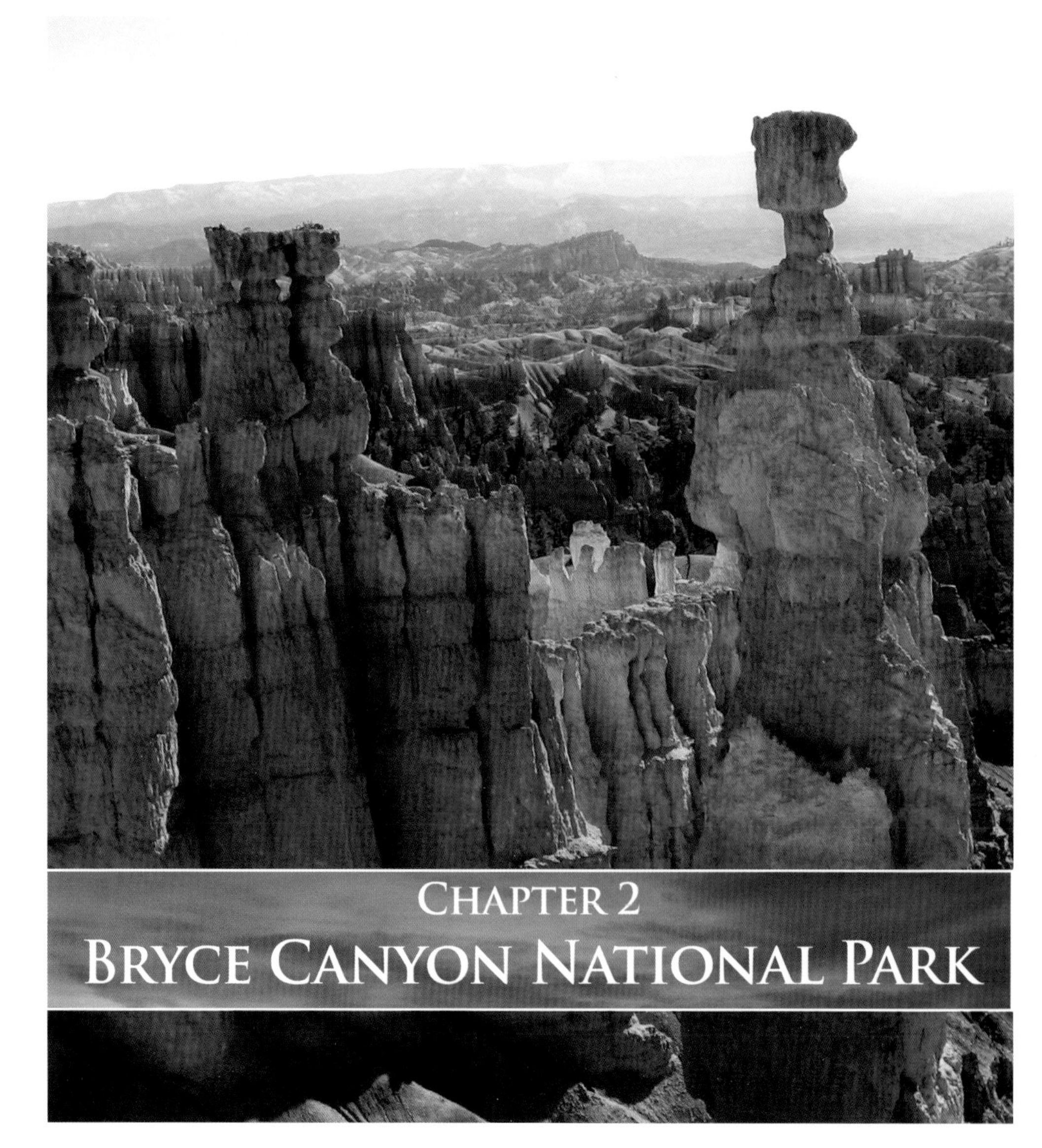

Chapter 2
Bryce Canyon National Park

Introduction

Bryce Canyon National Park was established in 1928 to preserve one of the most stunning landscapes on Earth. The brightly colored rock amphitheaters that beset the rim of Bryce Canyon were carved into the Eocene-age (approximately 50 Ma in the Paleogene Period) Claron Formation by processes of weathering and erosion. These "Pink Cliffs" can be viewed from great distances and represent the uppermost step of the "Grand Staircase." The amphitheaters house a labyrinth of rock chimneys, or hoodoos, which have captivated the imagination of people for centuries. The Native American Paiute tribe believed that the features of Bryce Canyon were the ruins of a great city and the rock chimneys represented its evil inhabitants turned to stone. Visitors penetrating the labyrinth of hoodoos and steep-walled canyons may well understand Ebenezer Bryce's famous description of Bryce Canyon as being "...a hell of a place to lose a cow." It has been said of Bryce Canyon that during the low-light hours of dawn and dusk it is difficult to take a bad photograph. Bryce Canyon National Park is a landscape of colorful contrast which inspires imaginations and invokes feelings of awe in all who venture to explore it!

Bryce Canyon at a Glance:

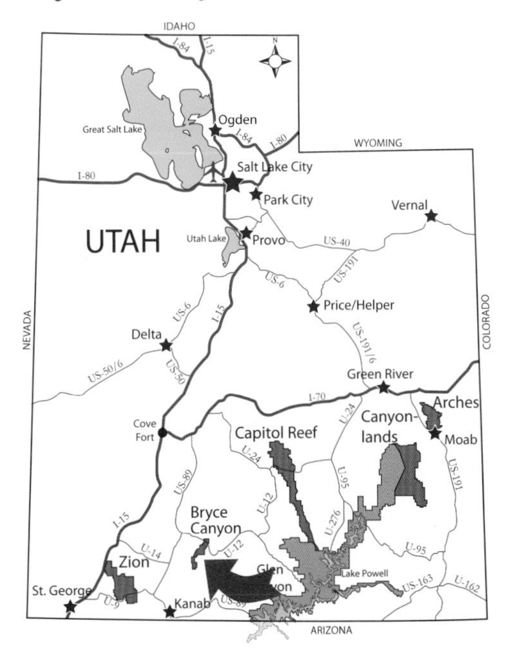

FAST FACTS:

Created as a National Monument:	1923
Given National Park Status:	1924 (Named "Utah National Park")
Named Bryce Canyon N.P.:	1928
Climate:	Semi-arid/Temperate
Land Area:	36,010 acres
Geological Province:	Colorado Plateau
Age of Exposed Bedrock:	Jurassic (200 Ma) - Neogene (20 Ma)
Hoodoo-forming Strata:	Claron Formation
Elevation:	7,600 to 9,100 feet (2,300 to 2,760 m) above sea-level
Rate of Rim Retreat:	1 ft (0.3 m)/100 years = 1.9 miles (3.1 km)/million years

How Were the Rocks Formed?

Figure 1 (right): *An inland sea referred to as the "Western Cretaceous Interior Seaway" deposited the Cretaceous-age strata in the area of Bryce Canyon.*

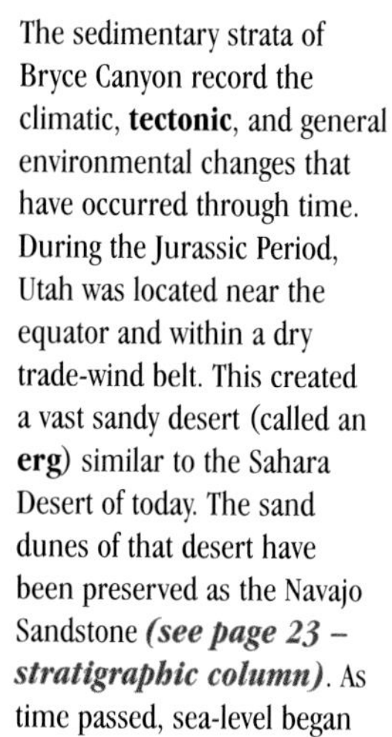

The sedimentary strata of Bryce Canyon record the climatic, **tectonic**, and general environmental changes that have occurred through time. During the Jurassic Period, Utah was located near the equator and within a dry trade-wind belt. This created a vast sandy desert (called an **erg**) similar to the Sahara Desert of today. The sand dunes of that desert have been preserved as the Navajo Sandstone ***(see page 23 – stratigraphic column)***. As time passed, sea-level began to rise, which created the Western Cretaceous Interior Seaway ***(Figure 1)***. Sea-level fluctuated over time in response to different climatic and tectonic changes. The Bryce Canyon area was situated along the margins of this seaway and recorded these changes in the rock record. The Dakota Formation was deposited in nearshore environments as the seaway first crossed (**transgressed**) the continent. As sea-level rose to its maximum depth, the Tropic Shale was deposited preserving numerous marine fossils. With continued sea-level fluctuation and final withdrawal, the Straight Cliffs, Wahweap, and Kaiparowits formations were deposited. These formations contain shoreline sandstones and coal from backshore swamps. Millions of years after the seaway withdrew (**regressed**) from the continent, the Claron Formation was deposited. The lower part (Pink Member) of the Claron represents deposits occurring on a broad plain including stream channel and deltaic sandstones, siltstones and pebble conglomerates. Mudstones formed from overbank flood deposits and ancient soils (paleosols) were created. The upper part (White Member) of the Claron Formation was deposited in broad lakes ***(Figure 2)***. As climatic changes effected the lake levels, different sediment was deposited: **limestone, dolomite,** and **carbonate rocks** in subaqueous environments, a mix of limestone/sandstone/mudstone (marl) in nearshore environments, and sandstone and pebble conglomerates in onshore coastal settings ***(Figure 3)***. Bed-by-bed rock variations helped to produce the famous **hoodoos** of Bryce Canyon National Park.

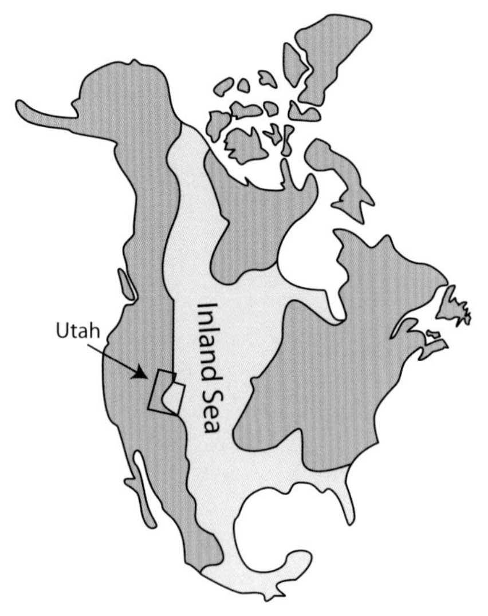

Figure 2 (right): *The 50 million year old (Eocene) intermontane lake system that created the Claron Formation in southern Utah. The red arrow displays the location of Bryce Canyon.*

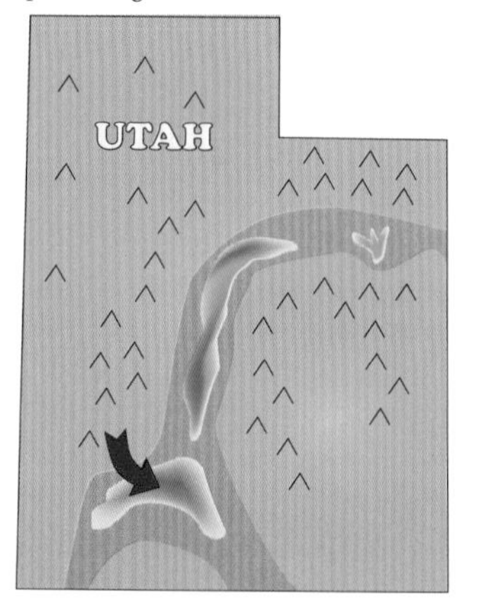

Figure 3 (below): *Modern photo of an active lakeshore environment, Lake Pukaki, New Zealand. This is similar to the environment in which the Claron Formation was deposited.*

How Were the Hoodoos Formed?

Figure 4 (right): An illustration which demonstrates how hoodoos are formed. Intersecting fractures allow running water to carve the hoodoos into the receding landscape.

Figure 5 (below): Photo of some hoodoos in Bryce Canyon. Notice how the hoodoos sit in rows. These rows are situated between prominent fractures that helped create the hoodoos.

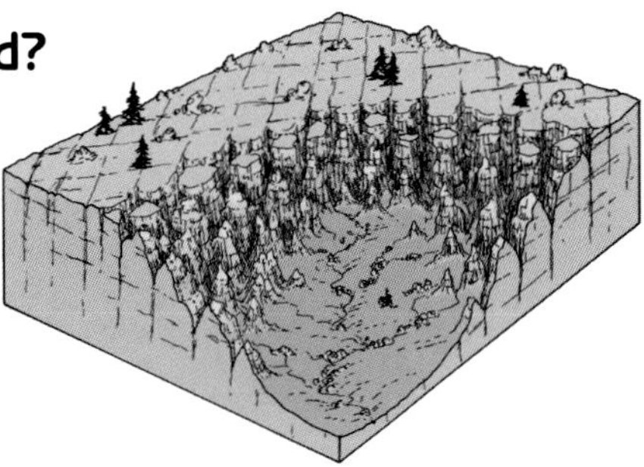

The hoodoos of Bryce Canyon are the product of multiple geologic processes. First, **plate tectonic forces** induced stresses on the buried rock strata during several geologic events occurring over the past 35+ million years. These events include: 1- the late stages of the **Laramide Orogeny** (a mountain building event approximately 85-35 Ma), 2- the uplift of the Colorado Plateau, and 3- present-day Basin and Range extension. Stresses produced systematic fractures within the Claron Formation of Bryce Canyon. Two fracture systems, one oriented to the northwest and another oriented to the northeast, intersect to

Figure 6 (above): This hoodoo is known as Thor's Hammer. It was named by European explorers who were travelling through the area. The stronger rock layers make the bulging parts, while weaker ones recess.

create a cross-hatched pattern of weakened rock ***(Figure 4)***. Second, as the Claron Formation was gradually exposed, physical and chemical weathering processes disaggregate the rock preferentially along the fracture sets. **Freeze-thaw cycles** occurring many times throughout the year accelerate these processes. Running water produced by summer thunderstorms carry the broken material downslope and expose new bedrock. Third, as this process continues, deep "slot" canyons and adjacent rock walls are produced. As intersecting fractures preferentially erode through the walls, columns of rock or **pinnacles** are formed ***(Figure 5)***. At the same time, the layers of sandstone, limestone, and mudstone comprising the column **differentially erode** creating curved recesses and bulging knobs ***(Figure 6)***. These combined processes produce the ghostly- and sometimes angelic-looking hoodoos.

Hoodoos like those found at Bryce Canyon are truly unique because of the processes involved in their creation. If tectonic forces had not created the intersecting fracture sets, the isolated and freestanding hoodoos would not have formed. If the climate were different, freeze thaw cycles and summer thunderstorms would not have been able to carve the hoodoos into the receding amphitheaters. If the strata of the Claron Formation were not so varied, they would not have formed the knobby balancing act that the hoodoos currently display ***(Figures 5 and 6)***. Given the delicate set of circumstances involved in creating these hoodoos, it is no surprise that they are so rare.

Arches, Bridges, and Erosion

Bryce Canyon also contains numerous **arches** and **natural bridges**. Arches form principally by chemical erosion along bedding contacts of a rock wall or **fin**. At Bryce Canyon these arches are essentially hoodoos that have not yet separated ***(Figure 7)***. Bridges form by running water tunneling through underlying weaker rock at a bedding contact.

Figure 7 (above): *The arches in this rock fin are formed in the same way as the hoodoos were formed. As erosion continues, the rock fin will be separated into hoodoos.*

The hoodoos, arches and bridges at Bryce Canyon are short-lived relative to geologic time. Weathering and erosion force the retreat of the canyon's rim at a rate of approximately 1 foot per century (0.3 m/100 years). Thus, in 1,000,000 years, the rim will retreat 1.9 miles (3.1 km), all the while destroying old hoodoos and creating new ones ***(Figure 8)***.

Figure 8 (right): *Weathering and erosion has exposed the roots of this tree.*

What Gives the Rocks their Color?

At times the rocks of Bryce Canyon appear to radiate various hues of pink, orange, and white ***(Figure 9)***. Color variation primarily results from the oxidation state and amount of the element **iron (Fe)** in the rock. Before the sediment was **lithified** (cemented into rock), it was deposited in nearshore and lake environments. The lake waters precipitated tiny crystals of **calcite** ($CaCO_3$) that accumulated on the lake bottom and eventually made limestone. The limestone is white in color because it contains little or no iron. The iron-bearing minerals and rock fragments that were deposited in oxygen-rich nearshore environments were oxidized into their ferric state (Fe_2O_3). This gave the sediment an orange color, like rust on a nail. Iron-bearing sediment deposited in oxygen-poor environments (often deep water) retains its ferrous state (FeO) and displays gray, green, or black colors (e.g. Tropic Shale).

Figure 9 (below): *The vibrant colors of Bryce Canyon can be attributed to the presence of oxidized iron (rust) in the rocks. Note the various bands of white running through the Claron Formation.*

Ancient **lava** flows, located north and west of the park, were rich in iron, but oxygen was unable to penetrate the molten **magma**. Therefore, the iron in the lava was not oxidized and the rock is black. These younger black **igneous** rocks stand in vivid contrast to the orange sedimentary rocks of the older Claron Formation at the Sevier Fault near Red Canyon west of the park. ***(Figure 10)***.

Figure 10 (below): *Photograph of the Sevier Fault. The black basaltic rocks on the left were originally higher than the pink rocks of the Claron Formation. Faulting has dropped the basalt so that it sits adjacent to the Claron Formation.*

What Is the Grand Staircase?

Sedimentary rocks of the Colorado Plateau record more than 1.6 billion years of Earth's history. Tectonic forces have lifted, **folded**, and tilted these strata, exposing them to high rates of erosion. Some rock units are more resistant to erosion than others. This is known as differential erosion. Stratified rocks that resist erosion form **cliffs**, while the rocks that are less resistant form **slopes**. When the entire rock column is slightly tilted by tectonic forces, differential erosion will create a series of alternating cliffs and slopes, like steps on a stairway ***(Figure 11)***. In southern Utah, the strata dip to the north, therefore the steps are especially visible as one looks north from the Grand Canyon area ***(Figure 12)***.

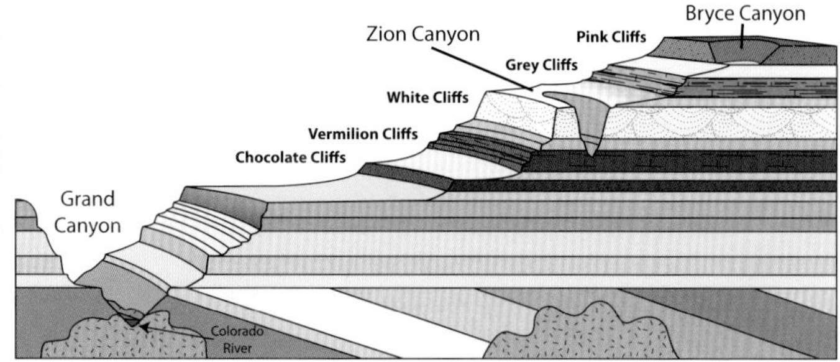

Figure 11 (right): *Illustration depicting the Grand Staircase. The Grand Staircase was created by erosion from the Colorado River and its tributaries. It is the host to several National Parks, including Zion National Park located in the White Cliffs. Bryce Canyon sits atop the Grand Staircase in the Pink Cliffs and is effectively the upper rim of the Colorado River drainage system.*

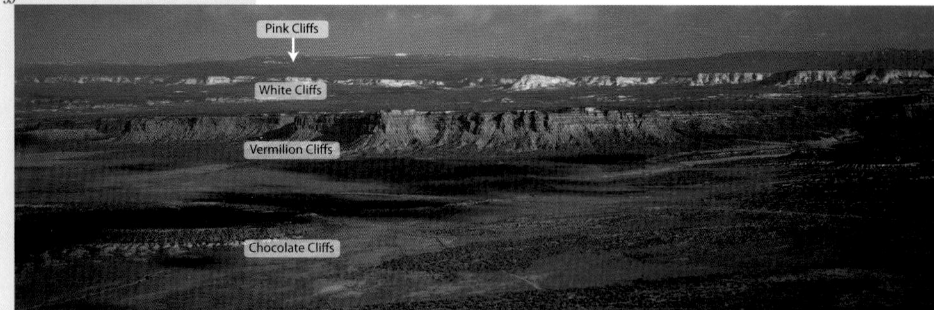

Figure 12 (below): *Northward view of the Grand Staircase from the Kaibab Plateau toward Bryce Canyon. The Cretaceous-age Gray Cliffs lie between the White and Pink Cliffs.*

Clarence Dutton first noticed these steps in the 1870's and referred to them as the "Great Geologic Stairway" which later became known as the "Grand Staircase." He named the various steps of the Grand Staircase according to their characteristic color. The top step of the Grand Staircase exposes the Claron Formation of Bryce Canyon and is known as the Pink Cliffs ***(Figure 13)***. The Grand Canyon sits at the bottom of the staircase approximately 100 miles (161 km) to the southeast. It is impressive to ponder the erosive power of the Colorado River and its tributaries as one envisions the amount of rock that has been carried away by the river to form the Grand Staircase (~11,000 vertical feet in the Grand Staircase alone). Indeed, Bryce Canyon can be considered the uppermost rim of the Grand Canyon.

Figure 13 (below): *These amphitheaters form the Pink Cliffs of the Grand Staircase. Note the basal Pink Member and upper White Member of the Claron Formation.*

Bedrock Strata of the Bryce Canyon National Park Area

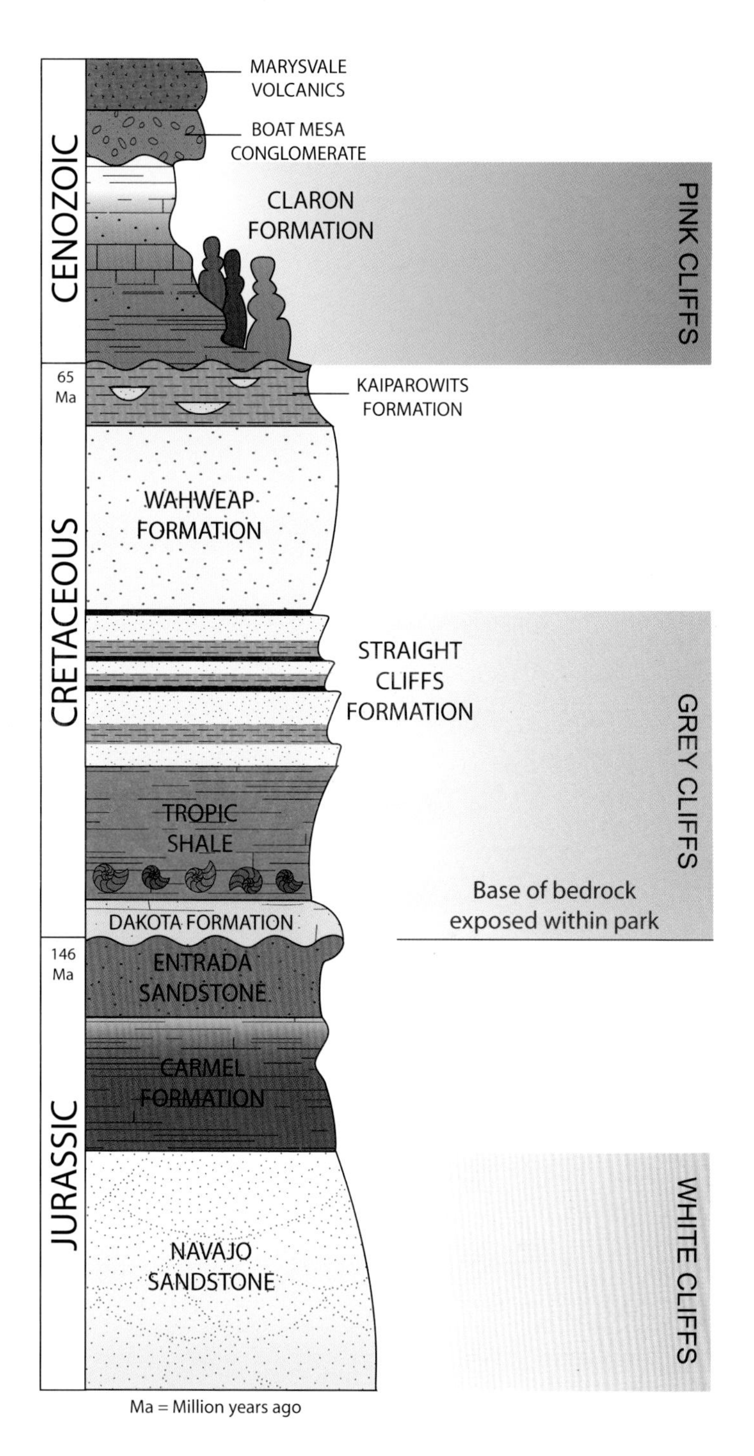

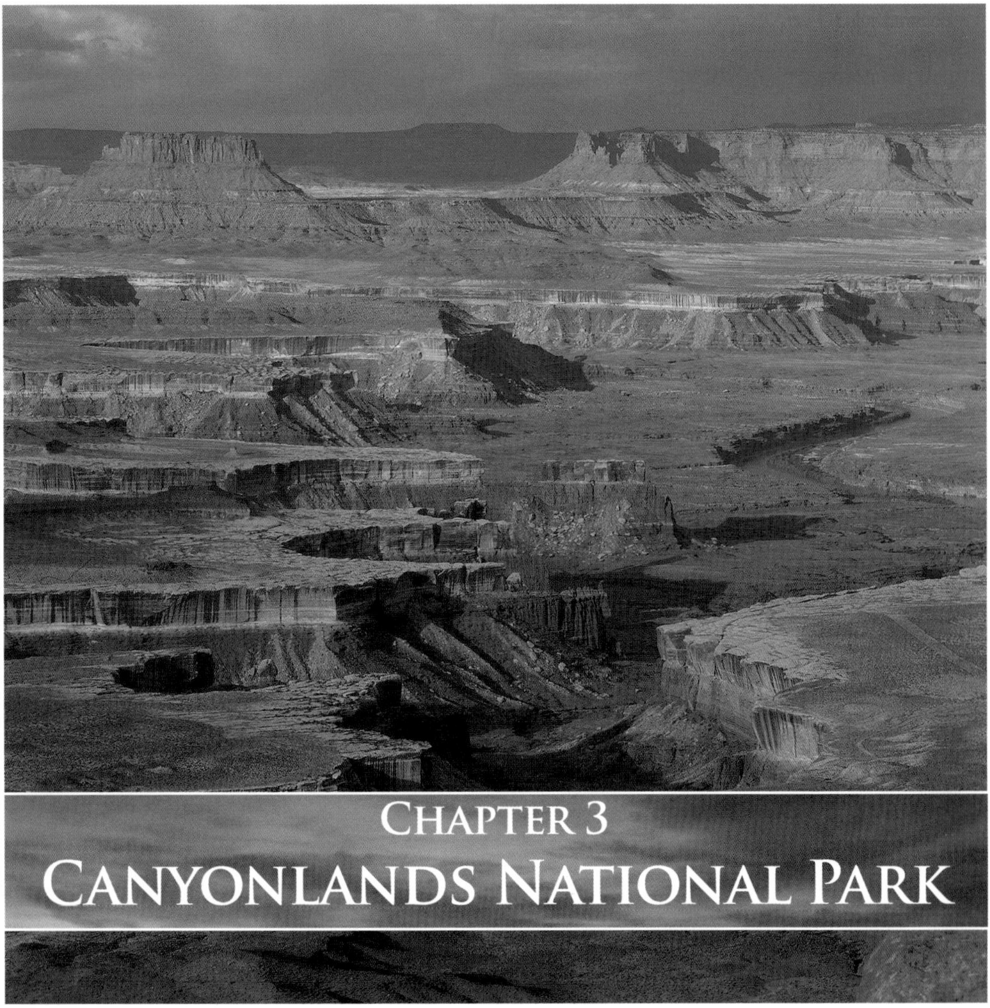

Chapter 3
Canyonlands National Park

Introduction

The centerpiece of Canyonlands National Park is the confluence of the Colorado and Green Rivers. The Y-shaped river system dissects the park into three equally enchanting regions. In the north, between the converging courses of the Green and Colorado Rivers lies the upland plateau district known as the "Island in the Sky." This district, home to the puzzling Upheaval Dome, is accessible by paved road, jeep trail, or on foot. Dead Horse Point State Park affords a beautiful view into the northeastern corner of Island in the Sky. The district located west of the Green and Colorado Rivers is known as "The Maze." The least accessible district, The Maze is a well-named labyrinth of ephemeral stream channels, gullies, and slot canyons carved into the Cedar Mesa Sandstone. Fins, arches, and spires sculpted from the Cedar Mesa also form the spectacular scenery of "The Needles" district, located in the southeast corner of the park. Sites worth seeing in this district include The Grabens, The Needles, and the confluence of the Green and Colorado. The singular beauty of the canyon country surrounding the Colorado and Green Rivers near their confluence was set aside as Canyonlands National Park in 1964 by President Lyndon B. Johnson. The park was expanded to its current size during the Nixon administration.

Canyonlands at a Glance:

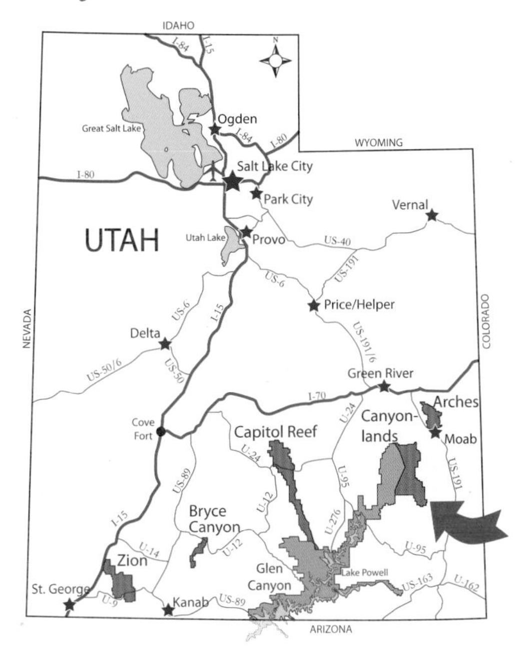

FAST FACTS:

Given National Park Status:	1964
Climate:	High Desert
Annual Precipitation:	8-9 inches (20-23 cm)
Land Area:	527 sqaure miles (1,365 square kilometers)
	337,570 acres (136,621 hectacres)
Geological Province:	Colorado Plateau
Age of Exposed Bedrock:	Pennsylvanian (330 Ma) to Jurassic (144 Ma)
Needle-forming Strata:	Cedar Mesa Sandstone
Elevation:	3,900 to 7,180 feet (1,189 to 2,188 meters)

Old Rocks, Young Canyons

Figure 1 (above): *Weathering (breakdown of bedrock) and erosion (the removal of weathered materials) by the Colorado River System, including its many tributraries (including the Green River), created the landscapes visible at Canyonlands National Park.*

Canyonlands is a showcase of sedimentary geology. The impressive variety of sedimentary **strata** exposed in the park provides the medium for the erosive power of the Colorado and Green Rivers to sculpt its characteristic canyons, plateaus, and needles ***(Figure 1)***. The modern scenery is a function of two processes: 1- deposition of sediment in oceans, streams, and deserts in the area that is now Canyonlands, and 2- **uplift** and **erosion** which followed millions of years later ***(Figure 2)***. Seventeen sedimentary rock formations are exposed in the Canyonlands National Park area ***(see stratigraphic column on page 31)***. These strata, which attain a thickness of thousands of feet, reveal a long history of deposition. The variety of rock types denotes sediment accumulation in a variety of **depositional environments**. Fossil-bearing **limestone**, salt, and marine sandstone of the Paradox, Honaker Trail, lower Cutler, White Rim, Black Box, and Moenkopi formations indicate deposition in shallow oceans and on adjacent tidal flats. Red and brown shale, siltstone, and sandstone of the Organ Rock, Chinle, and Kayenta formations was deposited by rivers and lakes that stretched across the area when the seas retreated. The major cliff-forming sandstone layers formed when Sahara-like dune fields (called **ergs**) covered the area, once in the Permian (Cedar Mesa Sandstone) and twice during the Jurassic (Wingate and Navajo Sandstones).

Figure 2 (below): *More than 100 million years after deposition, the strata were uplifted, allowing the Colorado River system to incise them. This created the labyrinth of canyons displayed in Canyonlands National Park.*

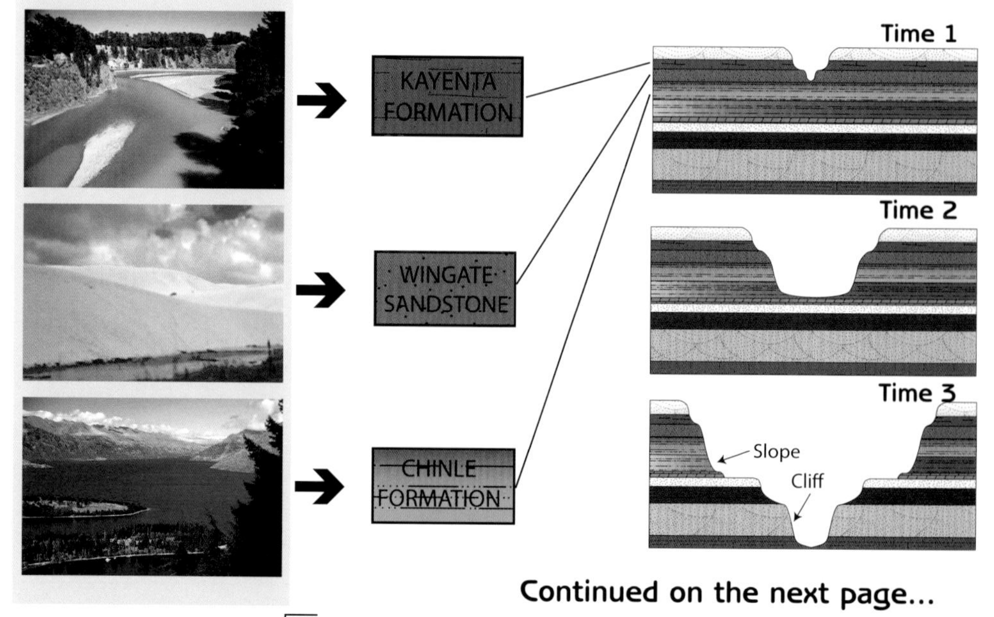

Continued on the next page...

Figure 3 (above): The "Turk's Head" area displays headward erosion. This is illustrated by the tributaries eroding back the canyon rims, giving them a scalloped appearance. ***Inset (right):*** *A dendritic (leaf-like) erosional pattern is displayed in The Needles district.*

Deposition effectively ceased 80 million years ago during the Cretaceous Period. The buried sedimentary layers were turned to rock (**lithified**) by precipitates from groundwater. Well into the Cenozoic Eon ***(see page 6)***, Canyonlands remained close to sea-level, drained by an ancestral Colorado River system. Then, **tectonic forces** initiated uplift of the Colorado Plateau. The Colorado and Green responded by down-cutting their channels and **headward eroding**, creating deep canyons, some reaching depths of 2000 feet ***(Figure 3)***. The rock that once occupied the space between the canyon walls (and more) has been eroded and transported downstream to the Gulf of California. The mud-laden waters of the Colorado indicate that headward erosion is still active, as the river strives to reach its **equilibrium profile** ***(Figure 4)***.

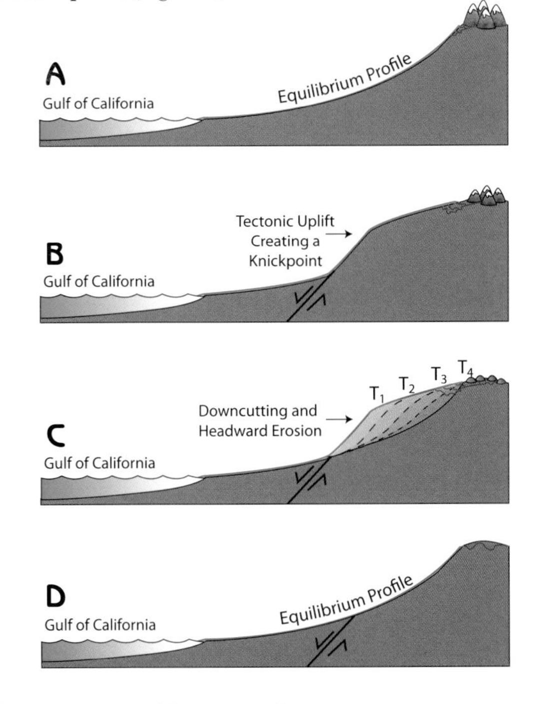

Figure 4 (right): *All rivers seek to establish an asymptotic (arcuate) equilibrium profile (A). The equilibrium profile is disrupted by uplift of the Colorado Plateau, which creates a knickpoint (B). The river system headward erodes through the knickpoint (C) to establish a new equilibrium profile (D). Canyonlands resulted from the river trying to re-establish equilibrium after the uplift of the Colorado Plateau.*

How Were "The Needles" Formed?

The Needles District permits park visitors to get "up close and personal" with the geology of Canyonlands National Park. Unlike The Maze, the Needles District is accessible by paved road and is crisscrossed by hiking trails and jeep trails. Permian strata in the district are likewise crisscrossed by a system of closely spaced **joints** (i.e. **systematic fractures**) ***(Figure 5)***. Rectangular blocks between the fractures have been reduced to columns and needles by **chemical** (e.g. dissolving) and **physical weathering** (e.g. **frost wedging** - the expansion of water as it freezes within fractures). Many of the columns and needles exist as freestanding landforms, bounded by joints that run parallel to the Colorado River.

Figure 5 (below): *The Needles were created by erosion along pre-existing, near parallel fractures in the rock.*

Alternating red/brown and yellow beds in the Permian Cedar Mesa Sandsone resulted from alternating deposition by **eolian** (windblown) and **fluvial** (river) processes. **Differential erosion** of less resistant interbeds contributes to the production of balanced rocks and arches such as the Wooden Shoe, which can be seen from the road near the Visitor Center ***(Figure 6)***.

Figure 6 (below): *The Needles display alternating red and yellow colored beds resulting from different depositional processes. These alternating beds also help create features such as "The Wooden Shoe"* ***(inset on right).***

How Were "The Grabens" Formed?

Figure 7 (above): These long linear valleys are referred to as The Grabens. Most valleys are created by stream erosion, but these were created by faulting and gravity gliding.

Figure 8 (right): Movement (see red arrows) of the Permian salt layer toward Cataract Canyon in essence provided lubrication for gravity to stretch and collapse the overlying rocks into fault-bounded "horsts" and "grabens". The stretching (extension) helped create the well developed fracture system in The Needles. Modified from Baars (2003).

The western third of the Needles District is home to The Grabens, an area characterized by elongate, flat-topped plateaus separated by long, flat-bottomed valleys ***(Figure 7)***. These features parallel the Colorado River. The valleys constitute down-dropped fault blocks (i.e. **grabens**) bounded on either side by uplifted **horst** blocks ***(Figure 8)***. The fault system responsible for The Grabens is related to westward gravity gliding and stretching of Permian strata situated above layers of Pennsylvanian salt. The salt layer undersgoes dissolution and deformation, which facilitates the stretching and gliding of the overlying **strata**. Gravity gliding is aided by the general westward tilt of the strata in the area. **Cataract** Canyon cuts across the western edge of the slump blocks, inducing continued movement of Permian strata toward Cataract Canyon. This gradual slumping constricts the canyon and creates the well known rapids (cataracts) for which it is named. As undercutting continues, so will glide-related widening of The Grabens.

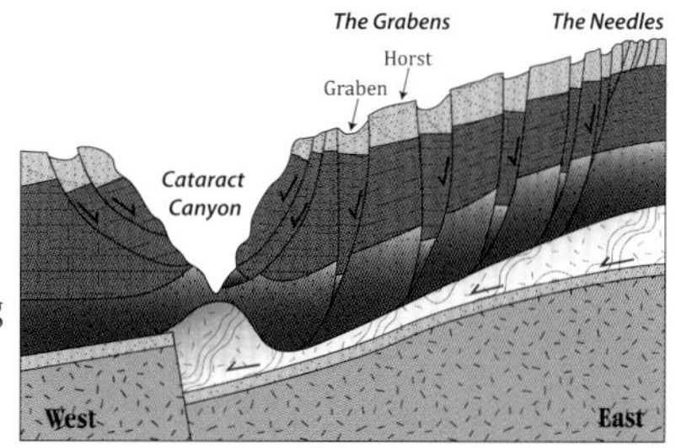

What is "The Maze"?

Figure 9 (below): Differential erosion along fractures within the Cedar Mesa Sandstone has created this unique landscape. Can you see why it is called The Maze?

The Maze is a tangled web of dry streambeds and passages between spires and knobs of Permian Cedar Mesa Sandstone ***(Figure 9)*** that cover approximately 119 square miles (308 square kilometers). This labyrinthine pattern is the result of two geological processes: 1- **uplift**, deformation, and fracturing of the bedrock and 2- **differential erosion** of the fractured Cedar Mesa Sandstone. As the Maze District was uplifted it was fractured along complementary sets of joints oriented roughly NE-SW and NW-SE. Fracturing was facilitated by flow of salt in the deeply buried Paradox Formation ***(see stratigraphic column on page 31)***. Although little, if any movement has occurred along these fractures, they provide a conduit for the infiltration of slightly acidic rain. The water dissolves the cement between the sand grains along these conduits. This results in enhanced erosion. Joints that have undergone this process for a long period of time have widened into linear stream valleys with the aid of flash floods. Narrow passages and slot canyons represent an early stage of the joint exploitation process.

What is "Upheaval Dome"?

An enigmatic, 1.6 mile-wide circular depression known as Upheaval Dome is located in the Island in the Sky District ***(Figure 10)***. The short hike from the road to the overlook is well worth the effort. Light-colored rocks in the core of the dome are bleached shales of the Moenkopi Formation mixed with sandstones injected from below. The layers of the overlying Chinle, Wingate, Kayenta, and Navajo formations dip away steeply in all directions from the central core, creating the large bullseye pattern that defines the structure.

Figure 10 (right): *Upheaval Dome, as seen from an aerial perspective. Inset photo was taken from a closer perspective and from an opposing view. Notice the bullseye pattern.*

The origin of the dome is controversial. The two leading explanations invoke terrestrial and extra-terrestial processes. One group provides compelling evidence that Upheaveal Dome is a **salt dome** ***(Figure 11)***. If so, the tilting of strata resulted from upward flow of a **diapir** (teardrop-shaped salt body) of Pennsylvanian-age salt. Such salt domes are common in the Gulf of Mexico and elsewhere.

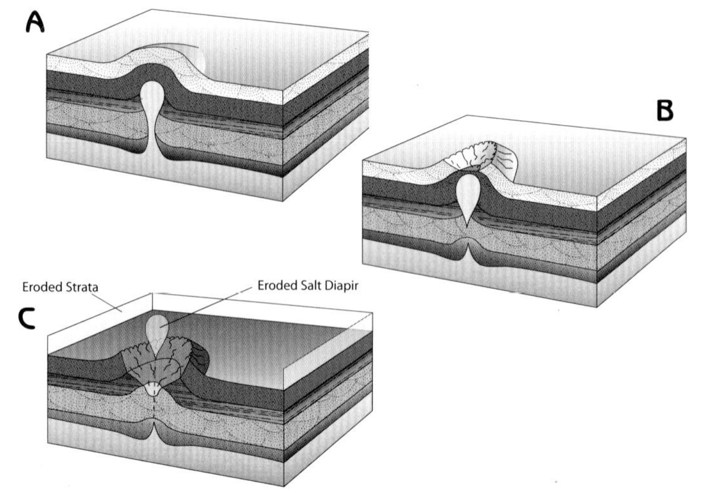

Figure 11 (right): *A) The salt diapir is less dense than the surrounding rock and begins to flow upward, causing the surface rocks to bend. B) The diapir pinches off as it continues its ascent. The uplift of the overlying rocks expose them to heightened rates of erosion. C) The pinched-off salt diapir, along with the overlying strata, are eroded away leaving behind Upheaval Dome.*

Another group has declared Upheaval Dome to be a **meteorite crater** ***(Figure 12)***. Proponents of this hypothesis claim that the circular stucture, intersecting sandstone layers, and strain patterns are best explained by a meteorite strike on Earth. This strike must have occured well after deposition of the strata. Another explanation could involve a combination of both processes, in which fractures from a meteorite impact provided a pathway for salt to move toward the surface. For the time being, the true origin of Upheaval Dome remains an enigma. Which interpretation do you believe?

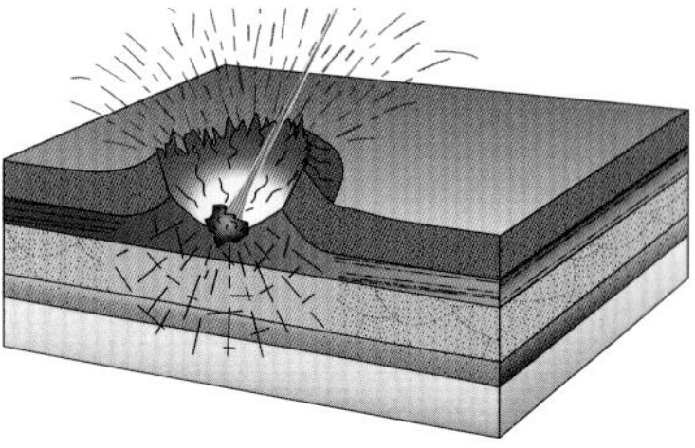

Figure 12 (right): *An impact crater could have also created Upheaval Dome.*

Bedrock Strata of Canyonlands National Park Area

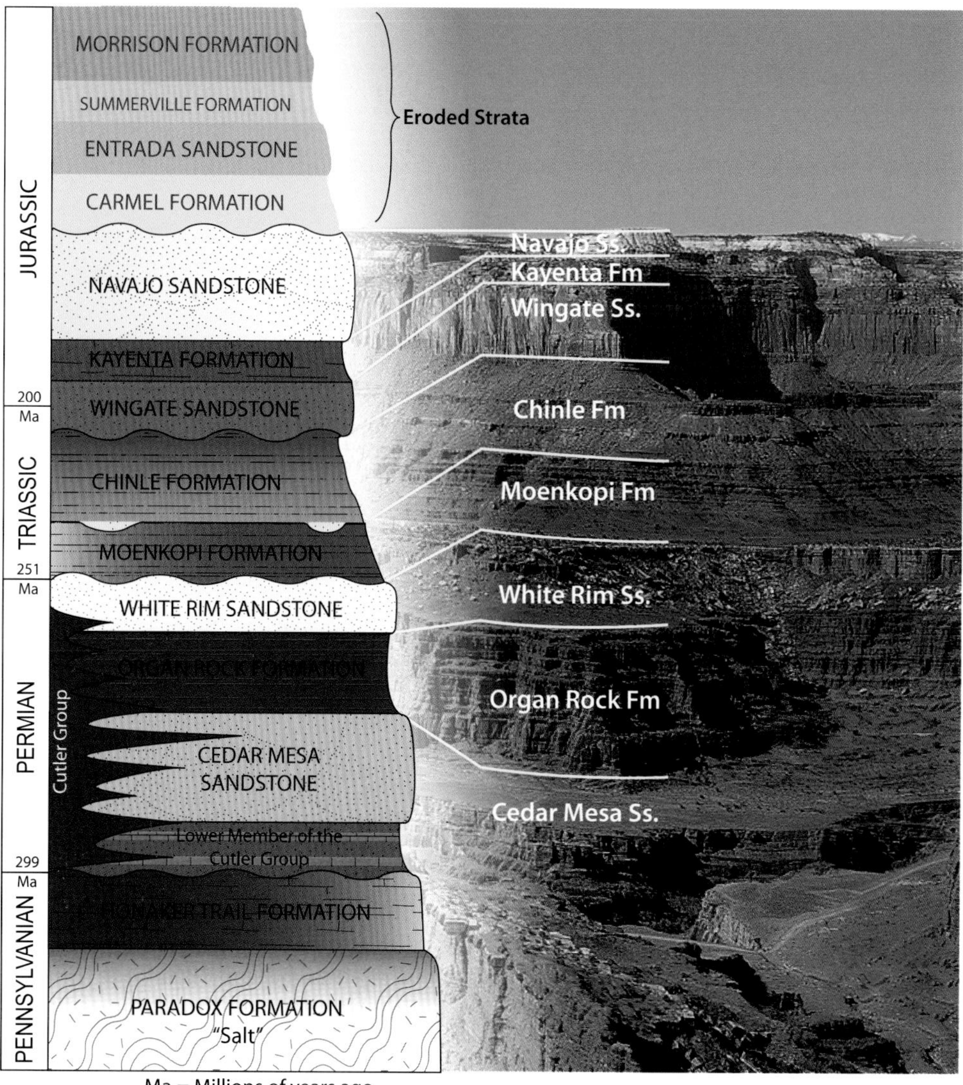

Ma = Millions of years ago

CHAPTER 4
CAPITOL REEF NATIONAL PARK

Introduction

Capitol Reef National Park was established in 1971 to preserve the desert landscape surrounding the Waterpocket Fold, a 90 mile-long monoclinal flexure of Earth's crust in the heart of the Colorado Plateau. Colorful sedimentary strata ranging from the Permian Period (275 million years old) to the Cretaceous Period (75 million years old) comprise the magnificent scenery of this high desert wonderland. The strata record deposition under ever-changing environmental conditions ranging from shallow ocean to river floodplain to desert dunes. Uplift and mountain building tectonic events (orogenies) then acted upon the strata creating folds, faults, and fractures. More recently, erosion involving wind and water has created the strike valleys, cathedrals, waterpockets, bridges, arches, and slot canyons observed in Capitol Reef National Park. To the casual hiker, pioneer historian, geoscientist, and "hard-core" backpacker, Captiol Reef National Park is an unheralded gem.

Capitol Reef at a Glance:

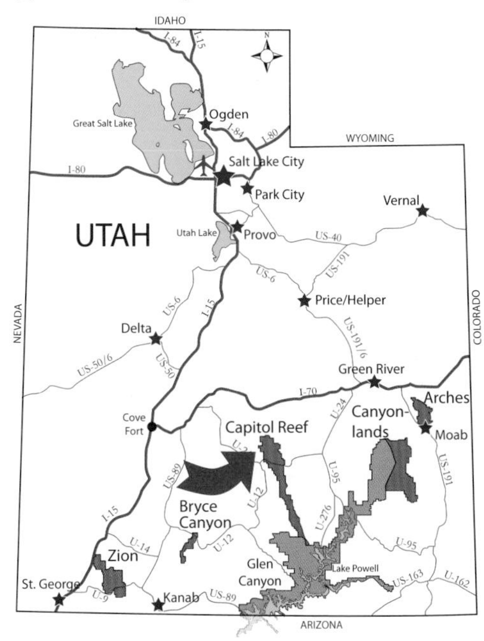

FAST FACTS:

Given National Park Status:	1971
Climate:	Arid (High Plateau Desert)
Annual Precipitation:	Approximately 7 inches/year (18 cm/year)
Land Area:	241,904 acres (97,895 hectacres)
Geological Province:	Colorado Plateau
Age of Exposed Bedrock:	Permian (275 Ma) to Cretaceous (74 Ma)
Elevation:	8,957 feet (2,730 m) to 3,878 feet (1,182 m)
Latitude:	38° 17' N (Visitor Center)
Longtitude:	111° 15' W (Visitor Center)

What is a Monocline?

A **monocline** is a fold in Earth's crust wherein one (mono) flank of the fold is steeply inclined (cline) and the other is nearly flat ***(Figure 1)***. The Waterpocket Fold, the geological centerpiece of Capitol Reef National Park, began forming between 70 and 50 million years ago in response to North America's westward movement over the Pacific Ocean Plate. **Tectonic** stresses passing far inland from the actual collision zone resulted in faulting and folding of the Capitol Reef area. Precambrian crystalline basement rocks located thousands of feet below Earth's surface responded to growing tectonic stresses by faulting. Precambrian rocks on the west side of the north-south trending fault plane were shoved upward approximately 7,000 feet. Overlying Paleozoic and Mesozoic ***(see page 6)*** sedimentary rock layers responded by passively draping the subterranean irregularity.

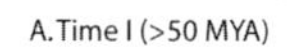

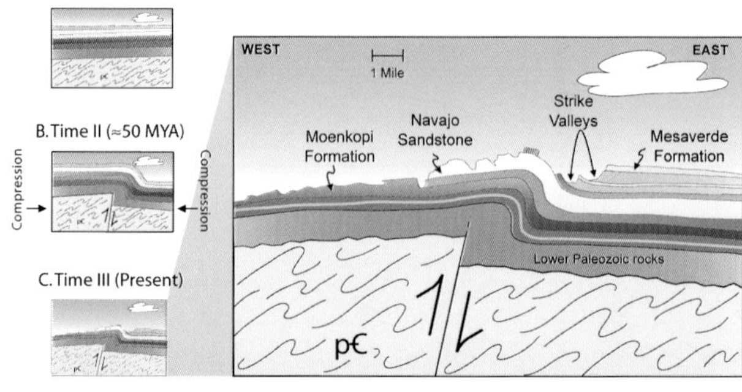

Figure 1 (right and below): Sequence of cross sections illustrating the development of the Waterpocket Fold through time. Compressional forces caused crystalline rocks of Precambrian age to undergo brittle deformation which resulted in a large reverse fault. The overlying sedimentary strata were more ductile (deformed like plastic) and draped over the underlying faulted rocks. Erosional processes at the surface have created the landforms of the park today. Aerial view looking north along the circle cliffs.

What are Waterpockets and How do they Form?

When sandstones are exposed for long periods of time, rainwater dissolves the cement between sand grains. This process proceeds rapidly where **cement** rims are composed of calcium carbonate. Atmospheric carbon dioxide mixing with rainwater creates a slightly acidic solution that is very effective at dissolving calcium carbonate. Where dissolution occurs, the sand grains are loosened and "blown away" by wind. Rainwater becomes concentrated in the resulting depressions, thus creating the **waterpockets** ***(Figure 2)***. In contrast, potholes and tanks are created by swirling currents that erode holes into the bedrock.

Figure 2 (right): Waterpockets like these are common on sandstone surfaces in the red rock country of the Colorado Plateau. In this photo, alkali-salt rimmed waterpockets attest to the evaporation of water which contained dissolved salts from the cement between sand grains. The grains are eventually blown away, creating surface depressions. The Waterpocket Fold derived its name from these features.

Why the name "Capitol Reef"?

Capitol Reef National Park derives its name from two features that captured the imaginations of early explorers. First, the maritime term "reef" was applied to seemingly impassible cliffs throughout the Colorado Plateau ***(Figures 1 and 4)***. Second, because the Navajo Sandstone in the Waterpocket Fold erodes into rounded landforms resembling the dome of the United States capitol building, ***(Figure 3)*** the area soon came to be distinguished as Capitol Reef.

Figure 3 (right): *View of Navajo Dome looking west. Located in numerous places in the park, domes such as these are found within the Navajo Sandsone. The Navajo Sandsone was created by deposition and burial of* ***quartz*** *sand in desert dunes that covered much of the Southwest during the Jurassic Period. The Navajo Sandstone contains some of the largest preserved dunes in the world* ***(inset photo)****. The sandy desert of the Navajo Sandstone is one of the many depositional systems represented in the strata of Capitol Reef.*

What is a Strike Valley?

Sedimentary **strata** on the east flank of the Waterpocket Fold are tilted steeply to the east (left edge of photo below). The eroded edges of the tilted strata are exposed along the entire north-south trend ("**strike**") of the Waterpocket Fold. As erosional processes etch the surface of the tilted strata, gray shale members of the Mancos Shale and mudstone-rich strata of the Entrada Sandstone are reduced to long, parallel **strike valleys** while the more resistant sandstones emerge as parallel ridges ***(Figure 4)***.

Figure 4 (below): *Aerial photograph of the "Strike Valley" looking south. Thick gray shale of the Cretaceous-age Mancos Shale is easily eroded thereby creating strike valleys. Resistant Jurassic sandstones (white and red) create the cliff line or "reef" of Capitol Reef.*

What makes the Rocks Red, Green, and Gray?

Color variation (red, green and gray) in sedimentary rock is usually related to the oxidation state of the element **iron** ***(Figure 5)***. Red rocks reflect oxidized environments which are commonly subaerially exposed. Gray and green rocks reflect reducing environments (subaqueous). White rocks contain little iron or no iron, perhaps as a result of leaching by fluids moving through the **pore spaces** of the rock ***(Figure 6)***. These fluids can be either groundwater or hydrocarbons such as oil or natural gas.

Figure 5 (right): *"The Castle" viewed from the Visitor Center. Blue gray mudstone of the Chinle Formation (low in photograph) was deposited in stagnant swamps and lakes on a low-lying coastal plain. These low oxygen environments produced iron compounds in their reduced state, giving the rock its gray color. The red rocks above and below, including the fractured Wingate Sandstone that creates "The Castle", were deposited in relatively oxygen-rich environments.*

Figure 6 (above): *Red flat-irons of the Carmel Formation abut the white tilted strata of the Navajo Sandstone. View is to the northwest from the Notom-Bullfrog Road in the southern part of the park.*

Why are there Black Boulders on the Red Rocks?

Approximately 25-20 million years ago, volcanoes poured out widespread andesitic and basaltic lava flows in south-central Utah. Remnants of these flows cap Boulder Mountain and Thousand Lakes Mountain located just west and north of the park. As these lava flows erode, black boulders are transported and deposited on top of the folded and eroded strata of the Waterpocket Fold ***(Figure 7)***. The downslope transport was the result of flash floods, debris flows, **alluvial** processes, and possibly glacial processes. It appears that boulder transport in the park was concentrated through the major existing drainages such as the Fremont River and Pleasant Creek. High on the bedrock above the present stream beds and adjacent to these drainages, geologists have mapped basalt-rich river terrace deposits. Some of these deposits are more than 500 feet above the elevation of the present stream bed.

Many of the black boulders have white patches on their surfaces. As boulders sit on the soil, alkali salts (caliche) precipitate on the underside of the boulders.

Figure 7 (below): *Basalt boulders perched on tilted sandstone bedrock are common in areas of the park where major stream drainages cut across the Waterpocket Fold. They are readily visible along Highway 24 in association with the Fremont River drainage. These boulders were derived from volcanic centers located west of Capitol Reef.*

Why do some Layers make Cliffs while other Layers make Slopes or Hills?

Strata undergo erosion as wind, sun, and especially water attack the exposed surface of the rock. Because there is natural variation in the size and mineralogical make-up of the constituent grains of different strata, some rocks are more easily broken down by these erosive agents. Landscapes result from this **differential erosion**. For example, the lunar-like surface created in the Bentonite Hills ***(Figure 8)*** results from the relatively rapid breakdown and erosion of the very fine-grained, clay-rich rock (smectitic mudstone).

In contrast, the Jurassic Wingate Sandstone is composed of **quartz** sand. Quartz is very resistant to erosional agents and that is why quartz-rich sandstones often create vertical cliffs. The long line of cliffs seen along the western escarpment of the park ***(see page 32)***, formed because resistant sandstones erode more slowly than the softer enveloping strata.

Figure 8 (below): *The "Bentonite Hills" are created within the Jurassic Morrison Formation. Regional volcanism was active at this time and ash falls were common to the alluvial plain of the Morrison. These ashes altered to smectite clays through geologic time and created the characteristics that result in rapid erosion and rolling topography.*

What is there to see in Cathedral Valley?

Cathedral Valley is best known for its spectacular rock towers or **monoliths**. These monoliths are erosional remnants of the Jurassic San Rafael Group. During retreat of valley walls, erosion is concentrated along **fracture** sets. This process isolates masses of less fractured rock (Entrada Sandstone). Where the Entrada is capped by the particularly resistant or unusually thick beds of Curtis sandstone, a monolith (temple or cathedral) may develop ***(Figure 9)***.

Evidence for **diapirs**, or plugs of ductile rock that have moved from the subsurface to the surface as domes, can be observed in at least two places in Cathedral Valley. "Glass Mountain" is composed of blocks of selenite (a colorless, transparent form of the mineral gypsum) ***(Figure 9, inset)***. It has been pushed to the surface from the underlying Carmel Formation as gypsum layers in the Carmel experienced increased burial pressure from overlying strata. If a gypsiferous diapir similar to Glass Mountain were to be dissolved by fresh groundwater, a deep circular hole in the floor of Cathedral Valley would result. This is how the "**Gypsum Sinkhole**" was created.

Figure 9 (below): *"Temple of the Sun" (middle) and "Temple of the Moon" (background) are two of the more spectacular monoliths observable in Cathedral Valley. Also in this area is "Glass Mountain" (foreground right), a selenite crystal diapir, or dome, that moved from the Carmel Formation in the subsurface to the Entrada Sandstone which floors Cathedral Valley.* ***(Inset on right)*** *Photograph illustrates the natural crystal form of selenite.*

Also observable in Cathedral Valley is evidence of a former volcanic field. Molten **magma** cut through the red sedimentary strata and cooled as black **igneous** rock ***(Figure 10)***. Igneous rocks injected vertically across the strata are called **dikes**. Igneous layers that parallel the sedimentary strata are called **sills**. These 4 million year old rocks could have served as the feeder pipes to surficial lava flows which have since been eroded away.

Figure 10 (right): *Black igneous rocks that cut through the red sedimentary strata are called dikes. These dikes, near the "Gypsum Sinkhole" in Cathedral Valley, are relatively resistant to surface erosion and stand out as "rock* ***fins****."*

Bedrock Strata of the Capitol Reef National Park Area

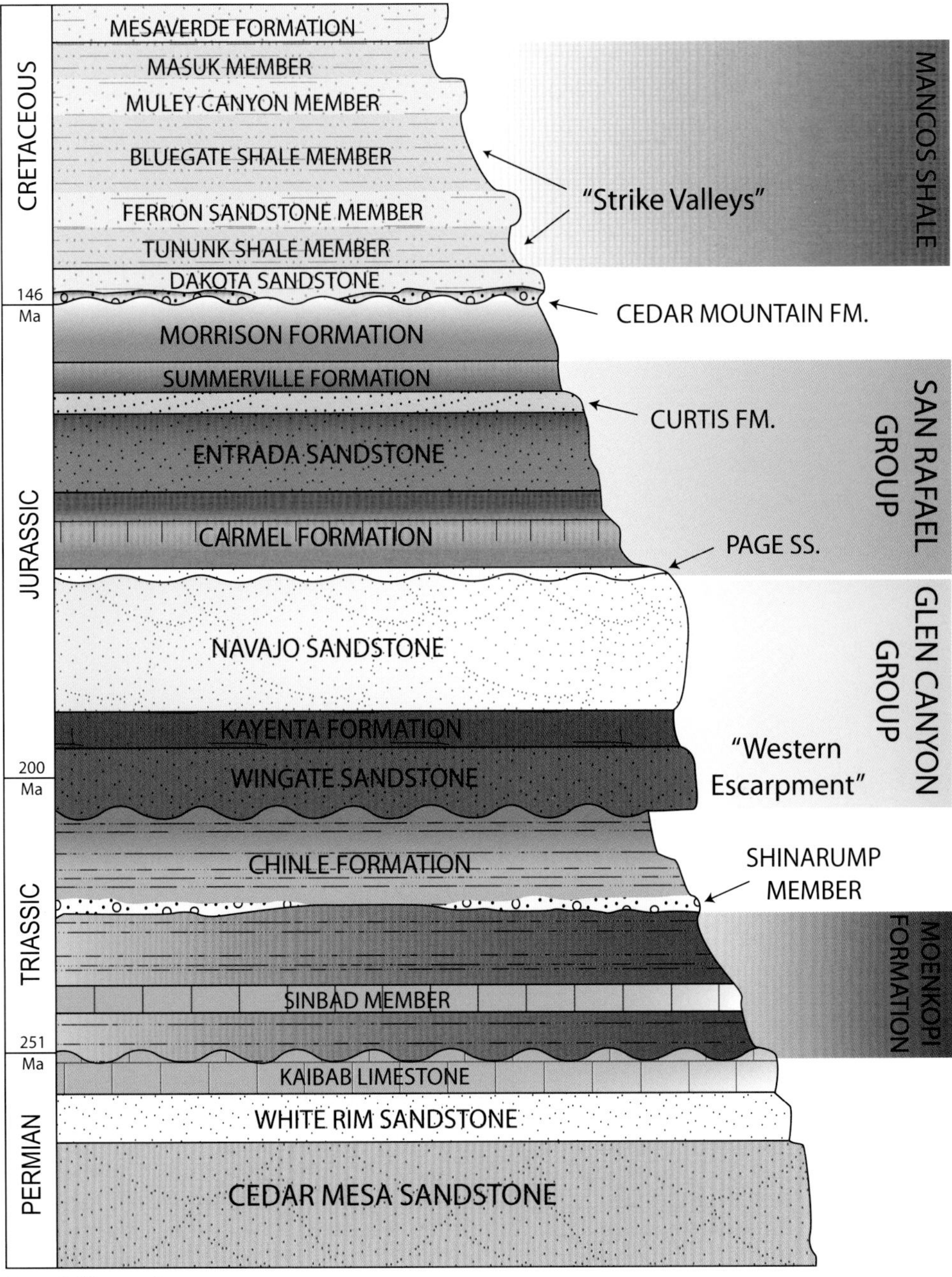

Ma = Millions of years ago

Chapter 5
Zion National Park

Introduction

Zion National Park, first established in 1919, is a testament to the erosive power of running water. The canyons, cliffs, temples, and towers of Zion were sculpted from the Navajo Sandstone and adjacent strata by the Virgin River and its tributaries over the past two million years *(Figure 1)*. At first glance, the labyrinth of canyons may appear random, but there exists a great deal of order to the landscape. Zion's landscape is controlled largely by a regional network of oriented fractures that are preferentially exploited by weathering and erosion. Triassic and Jurassic rocks exposed in the cliffs of Zion National Park tell the story of an ancient desert landscape where mud, silt, and wind-blown sand accumulated in large quantities during the age of dinosaurs.

The contrast of shadow and light on the land, the luminescent hues of rock formation set against the blue sky and green foliage, and the sheer scale of its cliffs and canyons make Zion a geological marvel and one of Utah's most popular national parks. This region was named "Zion" by Mormon pioneers who considered it a sanctuary where travelers could experience feelings of awe, immensity, and eternity.

Zion at a Glance:

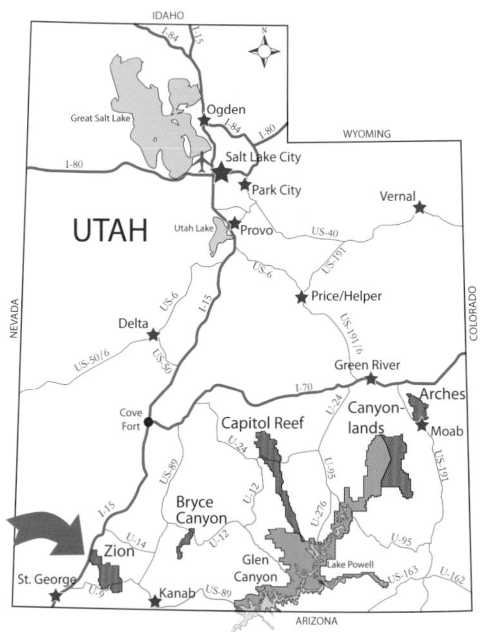

FAST FACTS:

Designated Mukuntuweap National Monument:	1909
Designated Zion National Park:	1919; additions in 1937 and 1956
Land Area:	146,597 acres (59,326 hectacres)
Elevation Range:	3666 to 8726 feet (1117 to 2660 m)
Geological Province:	Colorado Plateau
Latitude/Longitude:	~37° 12' N, ~112° 59' W at Visitor Center
Maximum Thickness of Navajo Sandstone:	~2200 feet (671 m)
Preserved Crossbedding:	Finest "fossilized" wind-deposited sand dunes in the world
Habitat Ranges:	Lowland Deserts (~14 inches precip/year [36 cm]) to Forested Highlands (~26 inches precip/year [67 cm])
Last Road-Damaging Landslide:	1995 lower part of Sand Bench beneath "The Sentinel" 1992 Balanced Rock Hills, Springdale

How was the Rock Formed?

***Figure 1 (above):** A westward aerial view of the majestic "West Temple," which is primarily composed of the Navajo Sandstone and is overlain by the Temple Cap Formation.*

***Figure 2 (right):** The Jurassic Navajo Sandstone was deposited in a sandy desert (erg) similar to the modern day Sahara Desert.*

The sedimentary rock layers (**strata**) of Zion National Park were formed by the accumulation of sediment within a variety of **depositional environments**. The Jurassic-aged Navajo Sandstone, which creates the prominent temples and towers of the park, was deposited in an ancient sandy desert (also called an **erg**) much like the Sahara Desert of today ***(Figure 2)***. During this time North America was only 10° to 30° north of the equator, sitting in Earth's dry, trade-winds belt. The erg eventually covered more than 100,000 mi^2 (260,000 km^2). This sand deposit reached its maximum thickness here at Zion National Park where it is ~2200 feet (671 m) thick. During deposition, the area was covered by large **dunes**, probably as high as 150 feet (46 m) or more. As **quartz** sand grains moved up the gentle windward side of a given dune, they would reach the crest and then avalanche down the steeper dune face, creating **crossbedding** ***(Figure 3)***. The compass direction of the dipping laminae that comprise the crossbedding indicates the direction that the Jurassic winds were blowing during deposition. The eastern portion of the park displays some of the most well preserved crossbedding in the world. Located stratigraphically beneath the Navajo Sandstone, the Kayenta and Moenave Formations were formed in rivers and ephemeral lake environments that eventually created finer-grained sandstone, siltstone, and mudstone. These formations are less resistant to weathering and erosion than is the Navajo Sandstone (see section on "Why Does the Canyon Width Vary?")

***Figure 3 (below):** Southwest view of the crossbedded and vertically fractured "Checkerboard Mesa" located in the eastern portion of the park. This ancient erg illustrates how avalanche faces of dunes get preserved as crossbeds **(insets)**. The predominant wind direction was from left to right.*

How did the Temples and Towers Form?

Figure 4 (right): "The Watchman" was formed as vertical fractures weathered, creating these pinnacles and towers. It is located on the east side of the park entrance.

The rock strata of the park were formed as buried and compacted sediments were **cemented** by underground fluids that turned the sediment to stone. This **lithification** process made the Navajo Sandstone particularly hard. During the Cenozoic Eon ***(see page 6)*** Utah began to rise in elevation due to the interaction of plate tectonic motions between the North American continental plate and portions of the Pacific oceanic plate. Stresses associated with this rise and later extension, formed **systematic fractures** and regional **faults** in the strata of southwest Utah ***(Figure 4)***. Over geologic time, erosion removed overlying rocks to expose the fractured rocks now observable in the park. These fractures created zones of weakness through which the tributaries of the Virgin River eroded deep canyons ***(Figure 5)***. Thus, many of the observable canyons which create the

Figure 5 (right): Systematic fractures observed in the red rocks (foreground) extend into the white rocks (background) to create narrow "slot" canyons.

temples and towers in Zion Canyon are **preferentially oriented** to the NNW and follow fracture sets ***(Figure 6)***. These erosional processes have removed thousands of feet of rock strata from the Colorado Plateau and Zion National Park. Erosion has primarily occurred in the past 2 million years. The eroded rock has been transported into the Colorado River drainage system via the Virgin River. The Colorado River transported this eroded material to its terminus in the Gulf of California.

Figure 6 (right): NNW-oriented fracture sets allow running water to carve V-shaped slot canyons which formed "The Court of the Patriarchs."

Why does the Canyon Width Vary?

The great width of Zion Canyon near the Visitor Center is in striking contrast to the slender slot canyons further upstream. Since all of the canyons in the park are carved by running water, one must look for other variables to explain their differences. Over time, water **downcuts** through the various layers of rock. Each layer of rock has different characteristics. Some are harder and more resistant to erosion than others. Harder rocks tend to form sheer cliffs and narrow canyons when incised by running water. As streams reach less-resistant rock, they erode more quickly and undercut the overlying resistant rock. This causes the overlying rock to collapse, which widens the canyon. The relationship between canyon geometry and hardness of the rock strata can be seen by investigating Zion Canyon. The slot canyon known as "The Narrows" was produced by the Virgin River cutting its way through the resistant Navajo Sandstone ***(Figure 7A)***. Further downstream, the water has sliced through the Navajo, and reached the less-resistant Kayenta Formation. These weaker rocks were no match for the powerful water, which undercut the Navajo by eroding away the weaker Kayenta. The effect of this can be seen by the widening of the canyon at the contact point between the Navajo and the Kayenta ***(Figure 7B)***. Even further downstream, the river reaches other weak layers such as the Moenave Formation, which further widens the canyon ***(Figure 7C)***.

Figure 7 (below): *Equal-scale topographic profiles illustrating the relationship between canyon geometry and stratigraphy. A) The canyon is narrow and v-shaped as the Virgin River cuts through the relatively resistant sandstone of the Navajo. B) The riverbed widens as it contacts the less-resistant Kayenta. Erosion of the Kayenta effectively undercuts support for the Navajo. Photo view is to the south. C) Upon cutting the relatively non-resistant Kayenta and Moenave, the river creates a wide valley with more gentle slopes. Aerial photo view is to the north.*

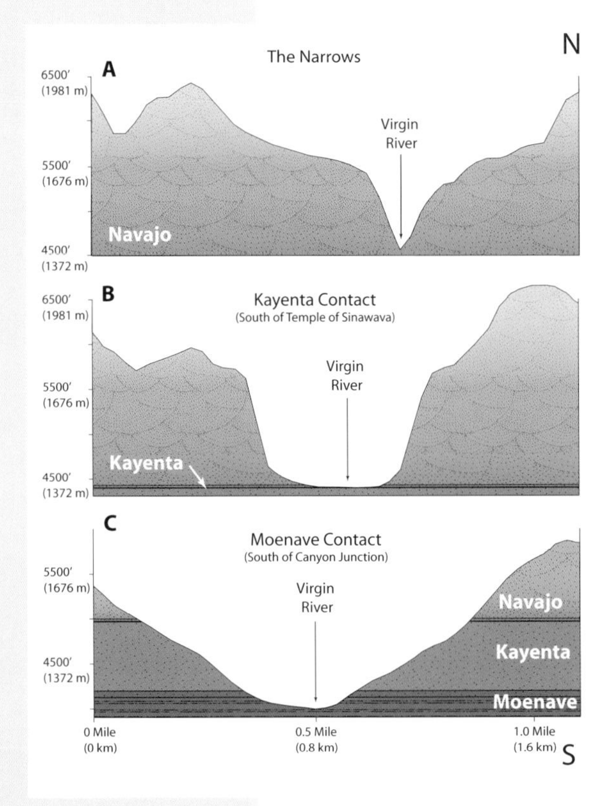

Why is there a Variety of Color on the Canyon Walls?

Figure 8 (above): *Westward view of the canyon wall just north of the "Towers of the Virgin." This photo illustrates the white (upper), pink (middle), and brown (lower) color variation in the Jurassic Navajo Sandstone. Note the weird contact between the pink and brown subunit that crosses the original primary stratification of the rock **(inset)**.*

Color variation on canyon walls can be: 1- an inherent characteristic of the sediment from which the rock formed, 2- created after the sediment was buried and lithified, or 3- "painted" on the surface after the rock is exposed.

Iron-bearing minerals comprise $<10\%$ of the Navajo Sandstone grains. If deposited under **subaerial** conditions, like a sandy desert, atmospheric oxygen will cause the iron to take on a red color, like rust on a knife blade ***(Figure 8)***. Therefore, before burial all of the Navajo was probably red. Today however, the upper part of the Navajo is white. After burial, acidic groundwater, possibly charged with natural gas, moved through the pore spaces of the rock and "**bleached**" the rock by dissolving the iron. Pink-colored rock below the white rock could represent the transition zone between gas-charged groundwater above and gas-depleted groundwater below. Vertical red streaks on canyon walls result from rain runoff as overlying red strata "bleeds" onto the canyon walls ***(Figure 9, see arrow)***. Areas and streaks of black, called **desert varnish**, result from precipitation of iron and manganese as groundwater and rain evaporates from the rock face. Clay minerals or microbiologic activity may aid this process.

Figure 9 (above): *The Sinawava Member of the Temple Cap Formation "bleeds" onto the face of the white Navajo Sandstone, staining "West Temple" with red streaks.*

Weeping Rock

The unique and fascinating spectacle known as "Weeping Rock" is one of the trademark features of Zion National Park. The water dripping from the alcove allows life to flourish. This precious water can be attributed to fractures that exist in the surrounding rock as well as the **pore space** of the Navajo Sandstone (i.e. porosity is up to 25% of the rock volume). These fractures joined forces with erosion to produce the alcove, above which lies a series of interconnected diagonal and vertical fractures that transmit rainwater downward towards a thin horizontal layer of **impermeable** mudstone ***(Figure 10)***. Upon reaching this obstruction, the water is forced to flow laterally towards the canyon wall. However, before arriving, the water encounters another set of vertical fractures, which conduct the water downward to the alcove.

Figure 10 (below): *Illustration and photo of the hydrologic pathway from rainwater to "Weeping Rock."*

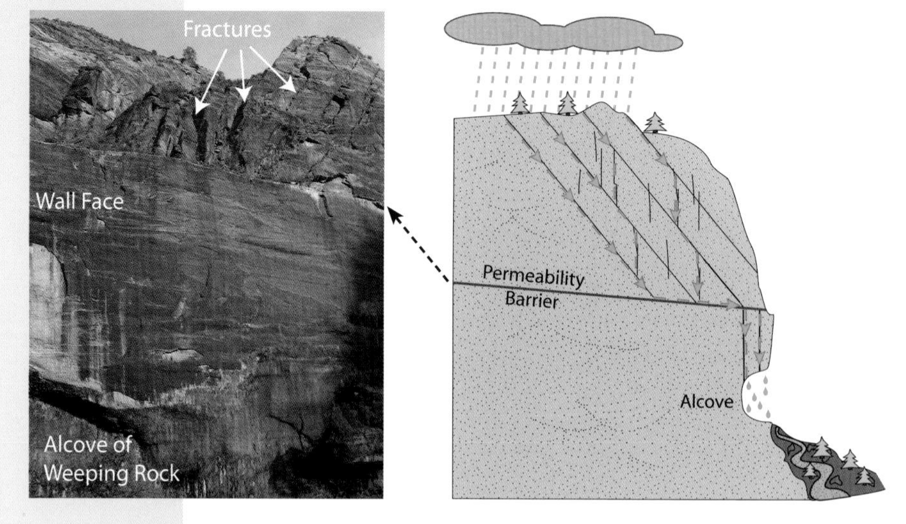

Landslides

Ancient and modern **landslides** are present in Zion National Park. Sand Bench is a large ancient landslide, located ½ mile north of Canyon Junction. On April 12, 1995, a significant portion of this landslide moved and dammed the Virgin River. This landslide involved approximately 110,000 cubic yards (84,000 m^3) of debris, enough to fill approximately 33 Olympic-size swimming pools.

Landslides consist of large blocks of rock that slide along **curved planes** of breakage ***(Figure 11)***. Several of these landslide features are observable in the 1992 Springdale Landslide, located just south of the park. Conditions that create landslides in the Zion area include: the various strengths of the rock units, amount of rainfall, slope angle, earthquakes, and gravity.

Figure 11 (right): *a) A conceptualized illustration of the Springdale Landslide and b) an associated aerial photograph. In 1992 the Springdale Landslide, resulting from a 5.8 magnitude earthquake on the nearby Hurricane Fault, destroyed three homes and blocked Highway 9. Most of the landslide is located in the Moenave Formation slipping along a shear surface within the Chinle Formation.*

Bedrock Strata of Zion National Park

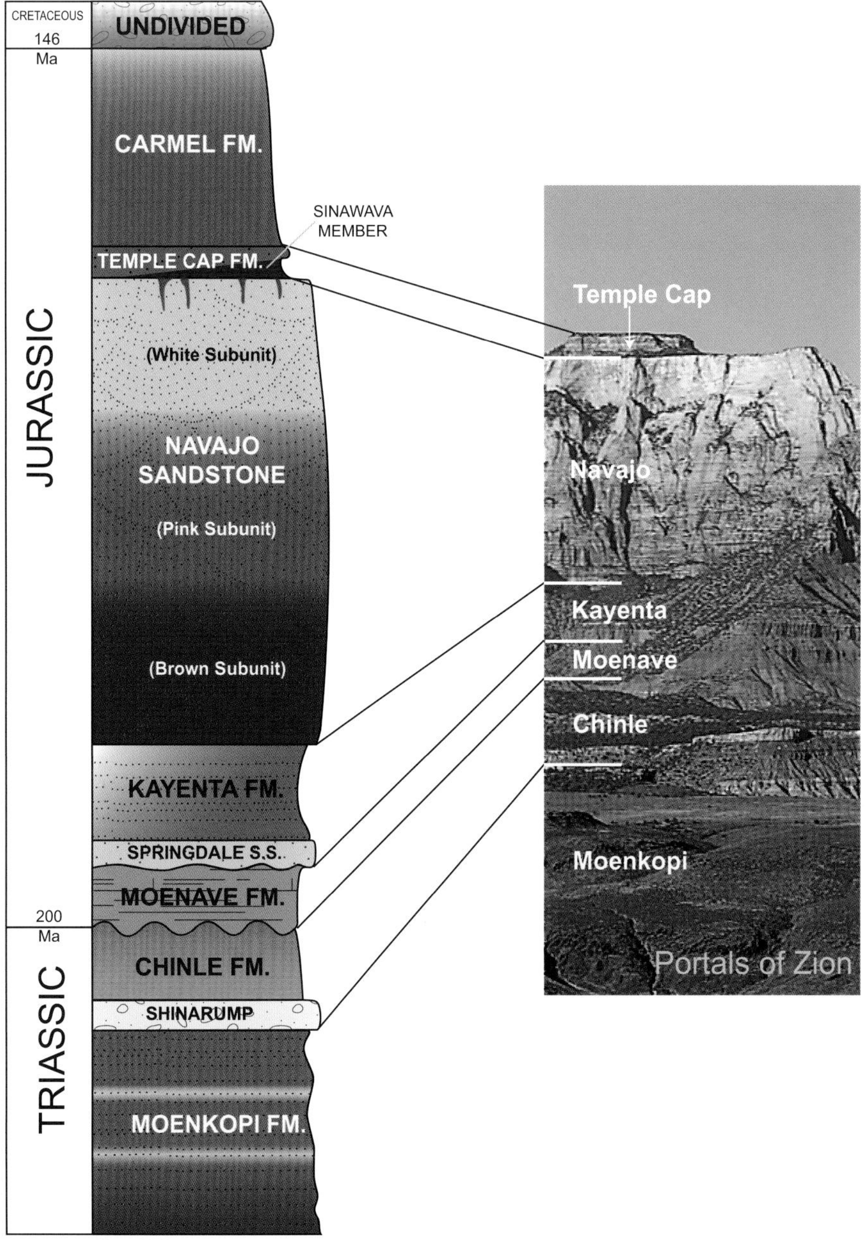

Ma = Million years ago

Chapter 6
Glen Canyon National Recreation Area

Introduction

Glen Canyon National Recreation Area (GCNRA) was created in 1972 to provide for public outdoor recreational use and enjoyment of Lake Powell and adjacent lands. The recreation area preserves scenic, scientific, and historic features. Named after explorer/geologist John Wesley Powell, Lake Powell is the second largest reservoir in the US. Lake Powell provides access to 96 side canyons along its 1960 miles of shoreline. Rock strata exposed along these shorelines include approximately 20 formations that span more than 300 million years of Earth history. Erosion and downcutting of the Colorado River and its tributaries through these bedrock formations give the area its spectacular color, form and beauty.

Glen Canyon at a Glance:

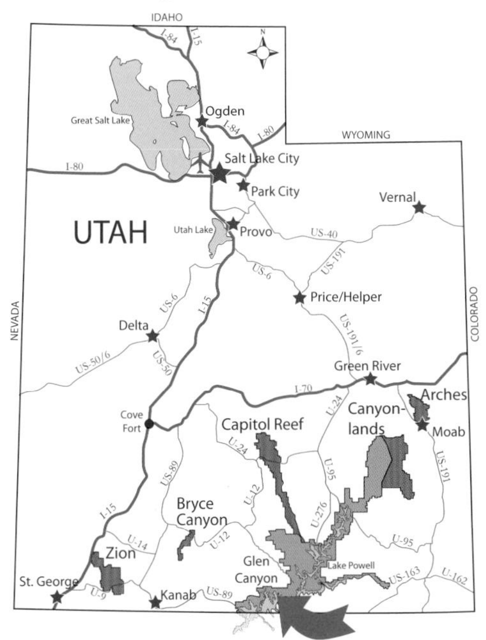

FAST FACTS:

Percent of GCNRA Covered by Lake Powell:	13%
Depth of Lake Powell at the Dam:	560 feet
Length of Lake Powell (at full pool):	186 miles
Elevation of Lake Powell:	3700 feet above sea level
Geological Province:	Colorado Plateau
Storage Capacity of Lake Powell:	27 million acre-feet
Hydroelectric Kilowatt Capacity:	1.3 million kilowatt hours
Annual Water Evaporation:	2.6 % of lake volume
Climate:	Arid (Desert)
Siltation Rate:	37,000 acre-feet/year
Time to Fill Dam with Sediment:	Widely speculated from 50 to 730 years

Why was a Dam created at Glen Canyon?

Figure 1 (above): The 710 foot high Glen Canyon Dam stores the waters of the Colorado River and its tributaries. In 1869, John Wesley Powell named the area Glen Cañyon because of the area's "...royal arches, mossy alcoves, deep, beautiful glens, and painted grottoes."

Construction of Glen Canyon Dam began in 1956 soon after receiving congressional approval. The dam was completed in 1963 and first filled to capacity in 1980. The dam provides water storage for the upper Colorado River basin states of Utah, Colorado and Wyoming. The dam was originally slated for construction at Echo Park, located upstream from Glen Canyon. Conservation groups objected to this site because a portion of Dinosaur National Monument would have been flooded. Glen Canyon was selected as a suitable alternative because the upstream area was undeveloped, the canyon was narrow and deep, and the Navajo Sandstone afforded a structurally sound foundation upon which to build the concrete dam ***(Figure 1)***.

How Did Rainbow Bridge Form?

Figure 2 (right): Rainbow Bridge is approximately 290 feet high and 275 feet across. At its apex, the rock span is more than 40 feet thick and 30 feet wide.

Through the past 1–2 million years, Bridge Creek cut through layer after layer of **strata**. Aided by existing **fracture** sets that were created during regional uplift, its **meanders** became well **entrenched**. Upon reaching the thick Jurassic Navajo Sandstone, the stage was set for the formation of Rainbow Bridge ***(Figure 2)***. Weak horizontal partings in the lower portion of the Navajo Sandstone allowed Bridge Creek to break through the meander neck before collapse of the overlying resistant rock. Ongoing erosion transformed the initial breach into the spectacular Rainbow Bridge. It is estimated that this occurred approximately 30,000 years ago. Geologists believe that Rainbow Bridge will continue to stand for thousands of years.

Canyon Cutting and Perched Gravel Deposits

Figure 3 (right): *Well rounded river gravels are found on hills and benches adjacent to Lake Powell. These cobbles and boulders, formed of very resistant rock types, represent former levels of the Colorado River.*

All of the landscapes observed in GCNRA are related to **downcutting** of the Colorado River and its tributaries. Several thousand feet of sedimentary rock have been removed by erosion in the last 3 to 5 million years, with up to 1200 feet of that being cut in the last 0.5 million years. This downcutting created the deeply incised Glen Canyon, which now forms Lake Powell upstream from the dam.

Gravel deposits are common along the beaches, ledges, and hills bordering Lake Powell. Two types can be distinguished by the careful observer; angular **alluvial gravels** and rounded **river gravels**. Alluvial gravels reflect accumulations of locally derived sand and gravel that were deposited in tributaries to the Colorado River. The angular shape indicates that these sedimentary particles did not travel far from their source areas. By contrast, the well rounded river gravels ***(Figure 3)*** attest to prolonged transport (tens to hundreds of miles) in the main trunk of the Colorado River. Both types of gravel deposits, now perched as high as 1400 (427 m) feet above the pre-dam river level, became "stranded" as the Colorado River changed its course through downcutting and meandering ***(Figure 4)***. Careful mapping of these gravels enables scientists to recreate the evolution of Lake Powell's scenic landscape.

Figure 4 (below): *Perched alluvial and river gravels help geologists decipher the history of downcutting along the Colorado River.*

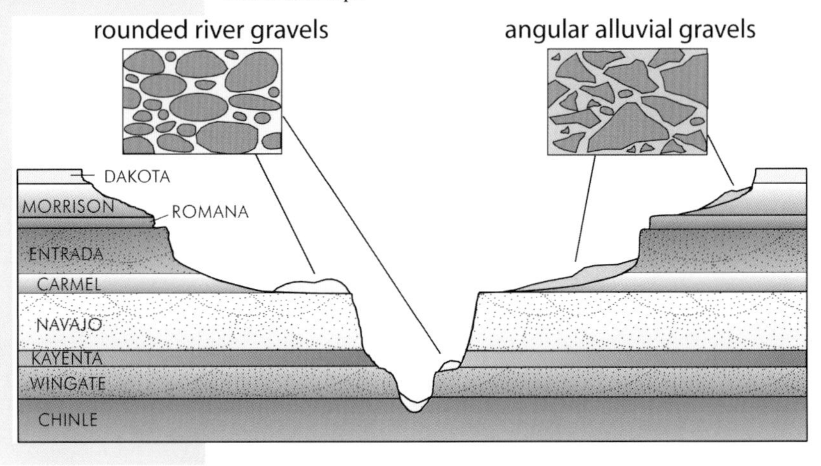

Entrenched Meanders - The Rincon

Geologists refer to loops, or "goosenecks," in a stream course as meanders. These result from the universal tendency of a river to migrate sideways when it flows across a flat-lying surface. If the flat-lying surface is uplifted, the stream will respond by downcutting into the underlying bedrock. Fractures in the bedrock may alter the shape of meanders which eventually become deeply entrenched into the uplifted surface ***(Figure 5)***.

The narrowest portion of the "peninsula" created by the meander loop is called the meander neck. With continued stream erosion, the meander neck narrows and eventually may experience "neck cutoff" wherein the stream flow abandons its pathway around the meander. This effectively straightens the stream's path, leaving behind a landform known as an **abandoned meander** ***(Figure 6)***.

Figure 5 (below): *Aerial photograph of several entrenched meanders at the confluence of the San Juan and Colorado Rivers. Note the narrowness of the meander necks.*

Figure 6 (right): *"The Rincon" is a classic example of an abandoned entrenched meander. After neck cutoff, the Colorado River shortened its course by more than 6 miles. The channel base of The Rincon is more than 500 feet above the modern (presently drowned) Colorado River channel base. The Rincon is one of many abandoned meanders along the Colorado River and its tributaries. Navajo Mountain, an igneous-cored uplift (**laccolith**) can be seen in the background.*

Geologic Hazards - Landslides and Rock Falls

Mass movement is the downslope transport of Earth materials under the force of gravity. Mass movement, in the form of **landslides** and **rockfalls**, commonly develops when fractured cliffs overlie more gently-sloping, soft, and unstable rocks ***(Figure 7)***. These occurrences can pose hazards to both campers and boaters. In GCNRA, the combination of fractured Wingate Sandstone above the clay-rich Chinle Formation creates a potentially hazardous situation ***(Figure 8)***. Landslides (usually in the form of slumps) are abundant in the Chinle Formation because it contains numerous beds of clay. This clay is an alteration product of volcanic ashes deposited during Chinle time. The largely impermeable clay collects water, allowing overlying rock to slide downslope quite easily ***(Figure 9)***.

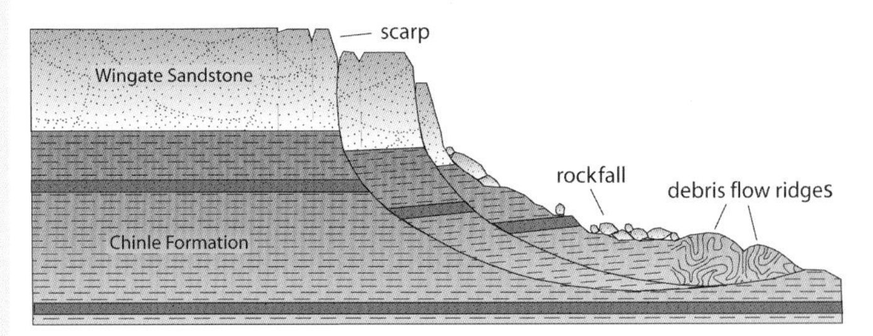

Figure 7 (right): *When erosion of softer rock (Chinle) undercuts fractured, cliff-forming strata (Wingate), rockfalls occur and blocks accumulate at the base of the cliff. Landslides occur as materials move downslope along a definite slippage surface. Landslide toes merge into debris flows.*

Figure 8 (right): *A modern rockfall, seen here at Good Hope Bay, occured when the fractured cliffs of the Wingate Sandstone were undercut by erosion of the underlying Chinle Formation.*

Figure 9 (below): *Characteristic features of massive landslides/slumps (right of mesa) occur within the Chinle Formation near "The Great Bend" of the San Juan River. The flat-topped mesa is supported by the cliff-forming Wingate Sandstone. Rockfalls have placed large blocks of Wingate Sandstone on top of the slumping Chinle Formation.*

Giant Weathering Pits, Injection Features, and Collapse Features - Padre Bay

Some of the most spectacular geologic features in GCNRA occur near Cookie Jar Butte in Padre Bay ***(Figure 10)***. These features include **giant weathering pits** (or holes), **injection domes** that cross-cut horizontal strata, and **collapse and fill features** ***(Figures 11 and 12)***. All of these features appear to have developed within the Jurassic Entrada Sandstone. Geologists cite several possible causes for the creation of the features including seismic triggering (earthquakes), meteorite impacts, or an ancient natural gas field. Whatever the cause, it is clear that **soft sediment deformation** was the prominent mode of transport for the sediment involved in the creation of the injection domes and associated withdrawal/collapse features. Soft sediment deformation occurs when unlithified, fluid-saturated sediment becomes buried. Due to overburden and some sort of triggering mechanism, the material begins to flow toward zones of lower pressure. As the sediment builds upward in some places, it withdraws from adjacent areas. These withdrawal zones cause overlying layers to collapse. Resulting depressions are later filled with sediment.

Figure 10 (below): *Aerial photograph of the peninsula adjacent to Cookie Jar Butte, Padre Bay. Note the numerous giant weathering pits (circular depressions) on left half of the peninsula.*

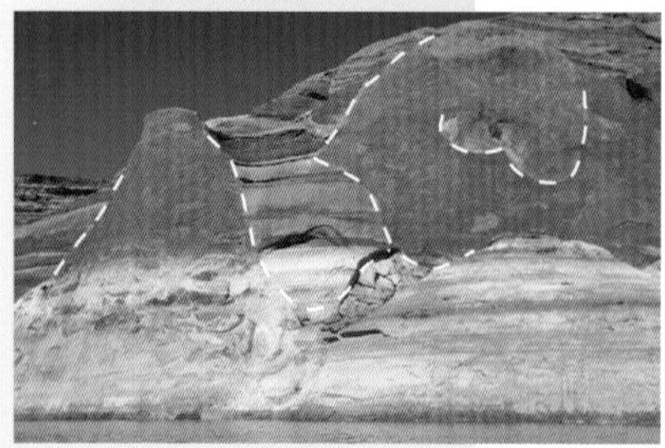

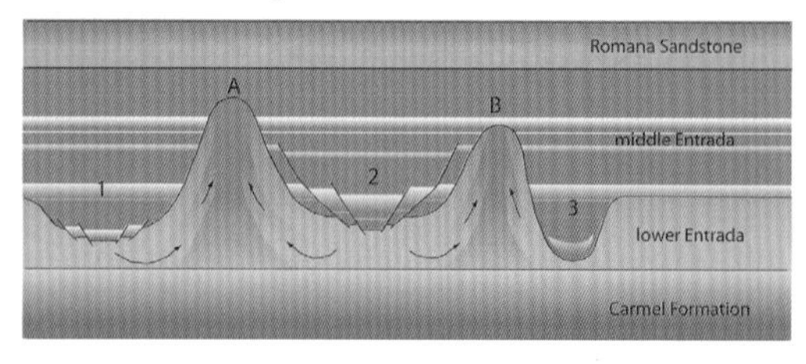

Figure 11 (above): *Photograph of injection domes illustrating the cross-cutting relationship between the vertically injected domes and the horizontal strata. Location is on the north wall across the bay from Cookie Jar Butte. Careful observation of these domes indicates that many resulted from multiple injections.*

Figure 12 (above right): *Illustration depicting injection domes (A and B) and adjacent withdrawal zones with associated collapse features (1, 2, and 3).*

Over geologic time the Entrada Sandstone became buried and lithified. Approximately 100 million years later it was uplifted. In the last several million years, the Colorado River and its tributaries stripped away the overlying strata. As the Entrada started to erode, some of the injected domes which were more highly **cemented** than the enclosing strata, weathered out in relief, creating mounds, pillars, and domes. Other injection domes were more poorly cemented than the surrounding rock. These domes weather into giant pits ***(Figure 13)***. Some of these pits drop vertically for more than 100 feet and pose a hazard to unwary hikers.

Figure 13 (right): *Giant weathering pits on the peninsula adjacent to Cookie Jar Butte.*

Bedrock Strata of the Glen Canyon National Recreation Area

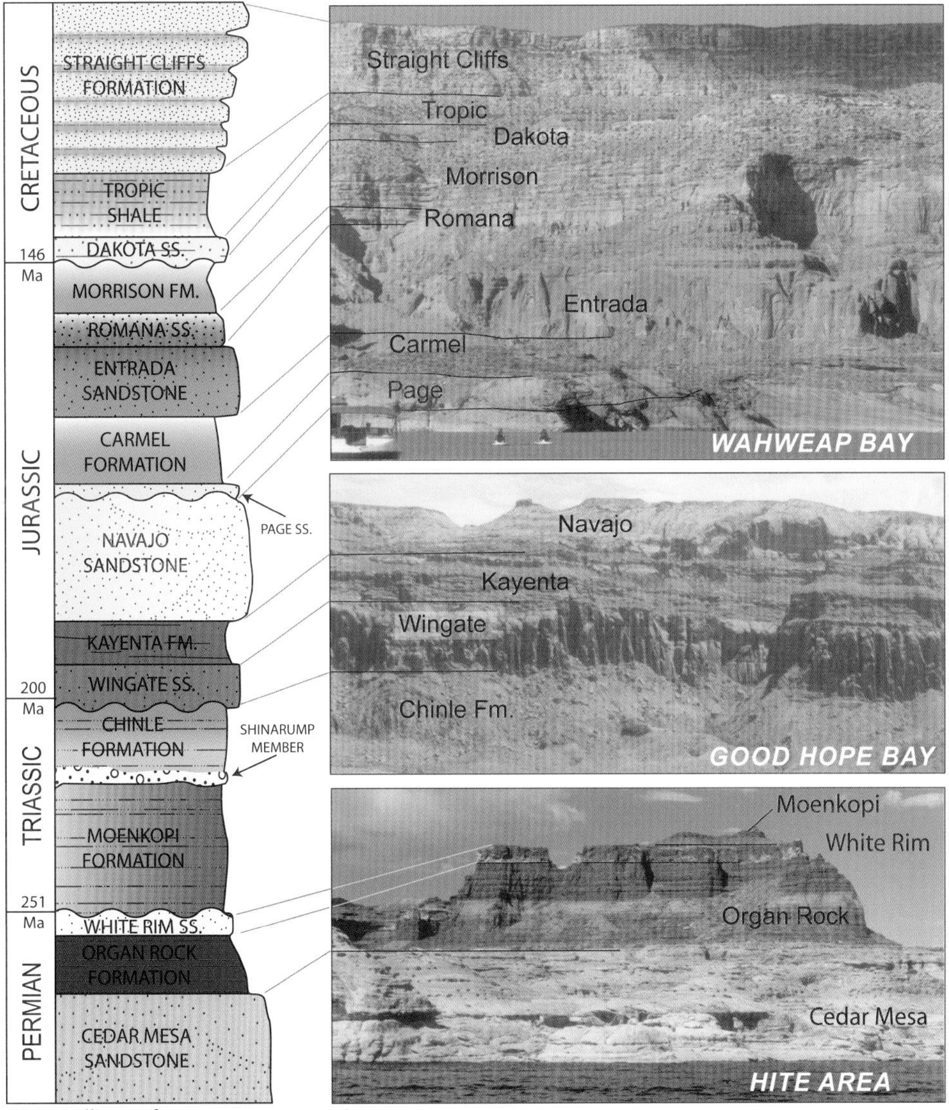

Ma = Millions of years ago

GLOSSARY:

The definition of each word in this glossary pertains to its common usage on the Colorado Plateau. After each definition we suggest other words that the reader may cross reference. Also, in parentheses, we reference the park(s) wherein the defined word may be observed (e.g. {B, CR}).

Abbreviations are as follows:
All = All parks display this feature or process in some form
A = Arches National Park
B = Bryce Canyon National Park
Can = Canyonlands National Park
CR = Capitol Reef National Park
G = Glen Canyon National Recreation Area
Z = Zion National Park

The following abbreviations are also used:
cf = compare/see also
e.g. = for example
etc. = and so forth
i.e. = that is

Abandoned Meander

A landform created by meandering streams and rivers. Erosion on the cutbank side of the meander creates an ever-exaggerated meander loop. Eventually the stream cuts itself off at the neck (neck cutoff) abandoning its former meander and creating a new course. The abandon meander may fill with water creating an oxbow lake. Cf: entrenched meander, meander. {G, Can}

Alluvial Gravels

Angular gravels that were deposited by flowing water relatively close to their original source (within several miles). The further they travel, the more rounded they become. More rounded gravels are referred to as "rounded river gravels". Cf: river gravels. {All}

Anticline

A crustal fold, usually induced by tectonic forces, in which the limbs dip away and downward from the fold axis (fold hinge or center-line). Anticlinally folded sedimentary rocks that have been eroded flat display successively older rocks toward the axis of the fold. Anticlines are capable of trapping large accumulations of petroleum in the subsurface. Cf: monocline, plate tectonic forces, syncline. {A,G}

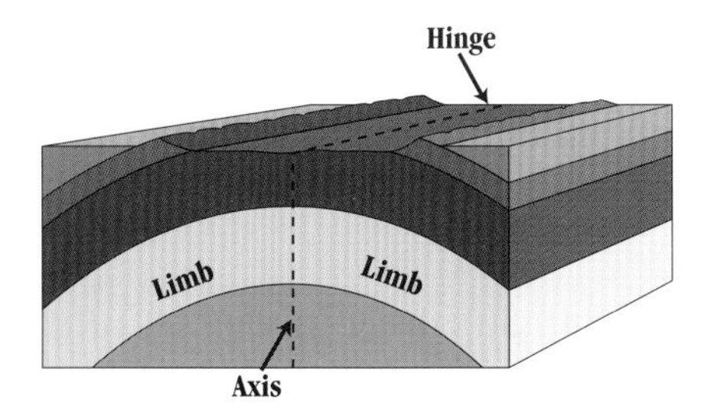

Arches

Arch-shaped landforms in excess of three feet (1 meter) in the horizontal or vertical direction. After a rock fin or protrusion is created by differential erosion along fractures, downward percolating groundwater can pond on impermeable layers such as bedding planes. The ponded water dissolves the cement around the overlying sandstone grains and eventually creates a hole through the rock fin. Cf: bedding planes, cement, differential erosion, natural bridges, fracture. {A, All}

Basalt

Commonly called lava, it is a dark colored extrusive igneous rock that is rich in iron (Fe) and poor in silica. Magma that flows on the surface of the Earth commonly creates basalt. Cf: igneous rock, iron, silica. {CR, B}

Base Level

The level below which a stream does not erode. For rivers terminating in the ocean, base level is essentially sea-level. Cf: equilibrium profile. {All}

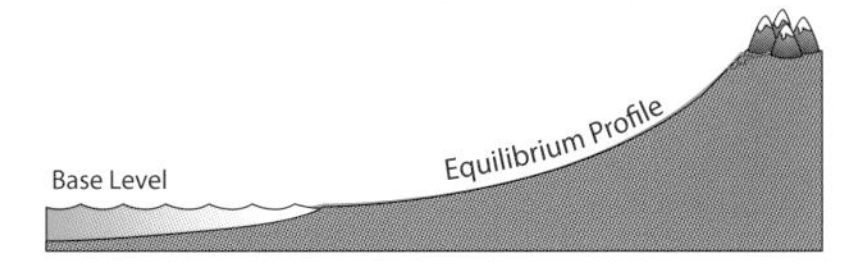

Bedding Plane

A surface or parting between different sedimentary rock layers. A bedding plane indicates some change in the depositional environment. Cf: depositional environment, strata. {All}

Bedrock

The continuous solid rock that underlies the soil and loose debris, and sometimes is exposed at the surface. An exposure of bedrock is called an outcrop. {All}

Bleaching

Any process by which a rock loses its color and becomes lighter. This can be accomplished by prolonged exposure to sunlight, or by fluids such as water and natural gas flowing through a rock and dissolving color-giving elements such as iron. Cf: iron, {Z, All)

Calcite

A mineral composed of calcium carbonate ($CaCO_3$). It is commonly precipitated from lakes and seawater. It is the primary mineral that makes up limestones but can fill pore spaces between sand grains to cement them into a sandstone. Cf: carbonate rock, cement, dolomite, limestone, pore space. {All}

Capitol

Capitol Reef National Park's name is in part derived from dome-shaped mountains of weathered sandstone that resemble the rotunda or cupola of many state capitol buildings. {CR}

Carbonate Rock

A rock composed mainly of carbonate minerals such as calcite and dolomite. Cf: calcite, dolomite, limestone. {B, CR}

Cataract

A waterfall or rapids in a river often created by a restriction in the width of the river's valley. {Can}

Cement

Minerals, usually formed from groundwater, that precipitate within the pore spaces of sediment. This binds the sediment together to create a sedimentary rock. This is one type of lithification. Cf: lithification, mineral, pore space, sediment. {All}

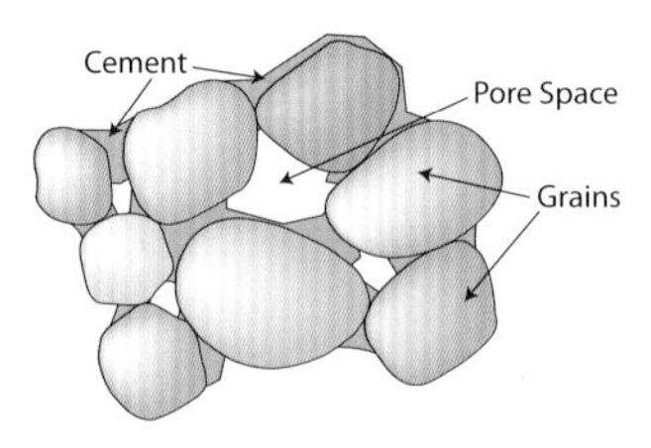

Chemical Weathering

Chemical reactions in which minerals and rocks are broken down, dissolved or otherwise altered by water or the atmosphere. Cf: physical weathering, weathering. {All}

Cliffs

Near vertical rock faces created by the erosion of strong, resistant rocks. Cliffs usually cannot be negotiated by humans without ropes and climbing aids. Contrast with less competent rocks, which tend to form slopes. Cf: erosion, slopes. {All}

Collapse and Fill Features

As sediment collects above underlying ductile beds (such as salt and gypsum), the ductile material tends to move away (withdraws) from the additional sediment load. Intervening strata collapse into the developing withdrawal zone, creating a topographic low on the surface which is, in turn, filled by sediments. Cf: injection domes, sediment, strata. {G}

Collapsed Salt-Cored Anticlines

If the crestal strata draping a salt wall becomes sufficiently thin (through erosion), rain and near surface groundwater can percolate down to the salt, and dissolve the upper portion of the salt wall. Overlying and flanking rocks can collapse into the developing void, creating a collapsed salt-cored anticline. Cf: anticline, erosion, salt-cored anticline, salt wall, strata. {A}

Cross-beds

Sand grains are pushed up the shallow, windward side of a given sand dune by the wind or, in the case of dunes in rivers, by the water current. Upon reaching the crest of the dune, they avalanche down the steep, lee face of the dune. These avalanches are preserved as "cross-beds" and can indicate the ancient flow direction. Cf: sand dune. {All}

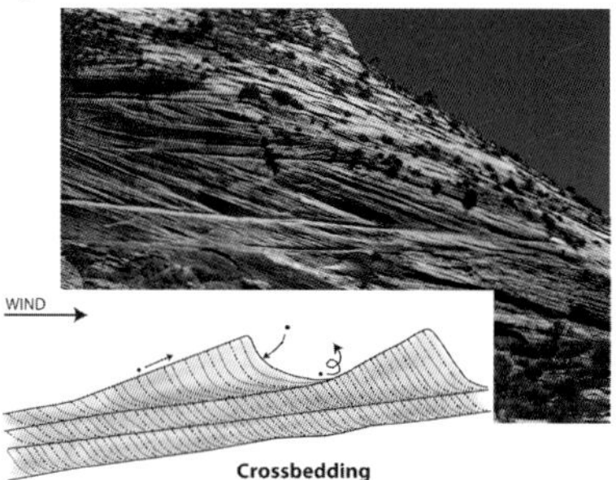

Curved Planes

Fault planes are often bowl-shaped in three dimensions. This causes the downward moving block to rotate backward toward the fault plane. Cf: fault, landslide. {Z, G}

Delta

A body of sediment deposited at the mouth of a river (e.g. Mississippi River delta). The sediment accumulates at a delta because the river has descended to base level and loses its energy and thus its ability to transport sediment. Cf: base level {B}

Depositional Environment

Sediment accumulates (deposits) in a variety of settings (environments) wherein they can be buried and lithified (i.e. turned to stone). These environments are usually places which have space to catch (accommodate) loose sediment. Examples include deltas, lakes, river valleys, continental shelves, etc. Cf: lithified. {All}

Desert Varnish

Stained areas and streaks of black on desert canyon walls and cliff faces. It is caused by precipitation of iron and manganese as groundwater and rain evaporate from the rock face. Clay minerals or microbiologic activity may aid in this process. Cf: cliffs, iron, minerals. {All}

Diapir

Plugs of ductile rock such as salt or gypsum that move through the subsurface towards the surface. Often these ascending rock bodies deform the adjacent strata. Cf: strata. {Can, A, CR, G}

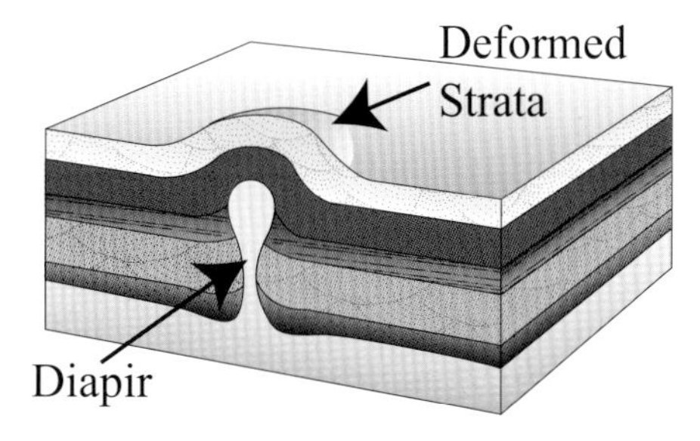

Differential Erosion

The erosional rate of different rocks varies due to their variable strengths. Weak and fractured rocks tend to erode faster forming slopes and valleys. Stronger rocks tend to resist erosion and form cliffs. Cf: cliffs, slopes. {All}

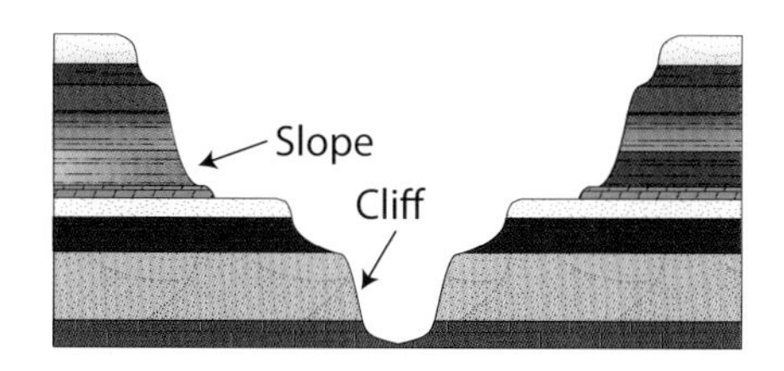

Dike

Usually an igneous rock that cuts across sedimentary strata. Over time, the surrounding sedimentary strata may differentially erode away, causing the dikes to stand out as rock fins. Cf: differential erosion, igneous rock, rock fin, sill, strata. {CR, B, Can}

Dip
The angle between horizontal and an inclined surface such as a bedding plane. Dip direction is the compass direction (azimuth) of the maximum dip angle. Dip and dip direction are common field measurements taken by geologists to measure folds and faults. Cf: bedding plane, fault, fold, strike. {All}

Dolomite
A carbonate mineral and rock that has been enriched in the element magnesium (CaMg(CO3)). In the semi-arid western states dolomite tends to be more resistant to erosion than shale and limestone. Cf: carbonate rock, limestone, mineral. {B, Can}

Downcut
The downward erosion (in elevation) of a river system. This often results in the creation of canyons. Rivers downcut in an effort to reach an ideal equilibrium profile and cease to downcut upon reaching base level. Cf: base level, equilibrium profile, erosion. (All}

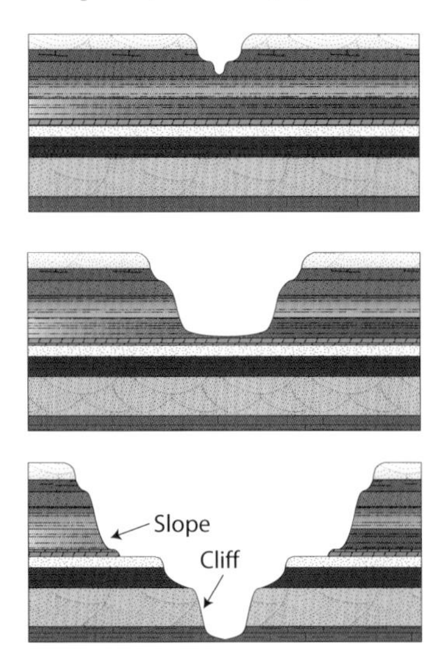

Drainage Divide
The ridge or highland area that separates adjacent drainage basins. {All}

Dune
See: Sand Dune

Entrenched Meander
A meander that has incised into the underlying bedrock usually because of local uplift or from a significant drop in base level. Cf: base level, meander, meander cutoff, uplift. {G, Can, CR}

Eolian
Refers to processes and areas (usually desert depositional environments) where wind is the primary mode of transport and deposition of sediment. Cf: depositional environments, sediment. {Z, A, Can, CR, G}

Equilibrium Profile
River systems strive to achieve an asymptotic (arcuate) stream profile from their terminus to their headwaters. This implies that the river system is near horizontal at the mouth, and becomes exponentially steeper upstream. When this equilibrium profile becomes disrupted, the river system adjusts to re-establish the desired profile. This either involves erosion, if the stream has been uplifted or base level has dropped, or deposition, if the stream has been lowered or base level has risen. Cf: base level, erosion. {Can, All}

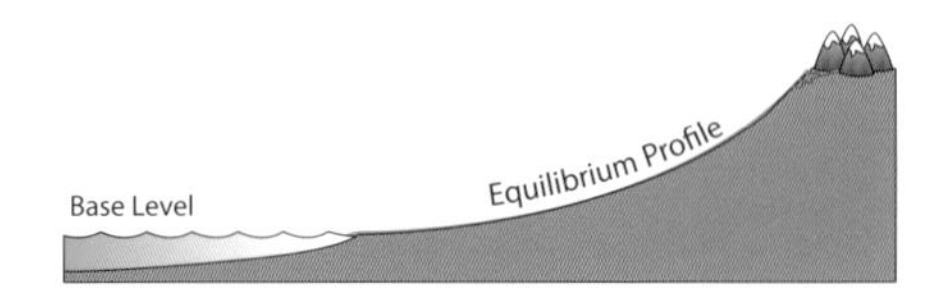

Erg
Refers to a specific type of eolian depositional environment: A vast area covered in sand dunes (e.g. Africa's Sahara Desert). It is often described as being a "sea of sand". Cf: depositional environment, eolian, sand dune. {Z, A, Can, CR, G}

Erosion
The removal and transport of sediments, after they have been weathered from the rock. Water is the most prolific agent of erosion; however, other agents of erosion include wind, ice, and gravity. Cf: weathering. {All}

Fault
A fracture which has experienced displacement (either vertical or lateral). Faults usually result from tectonic-induced stresses that are either compressional, extensional, or shear (lateral or translational). A variety of fault types can exist including: normal, reverse, strike-slip (lateral or translational), or thrust faults. Cf: fracture, joint. {All}

Fault Block
A rock body that is fault bounded and is often described as the footwall block or the hanging wall block. Cf: fault, fault scarp. {All}

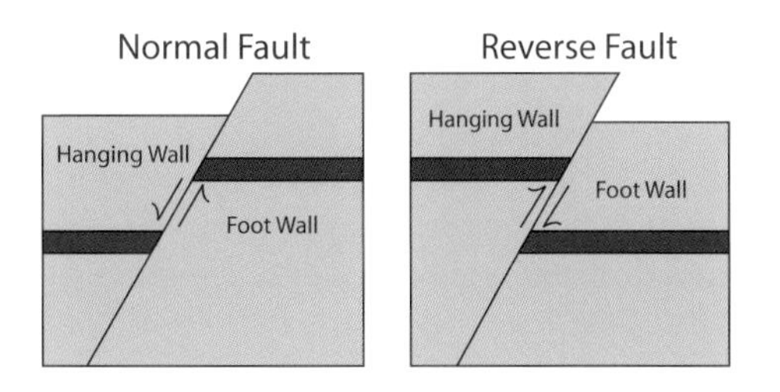

Fault Scarp
A cliff or ledge produced by fault movement. The detachment area of the fault block from the bedrock produces a steep rock face. Cf: cliffs, curved planes, fault, fault block. {Z, G, CR}

Fin
An isolated wall of rock formed by erosion along fractures or systematic fracture sets. Also produced by differential erosion around a dike. Cf: differential erosion, dike, systematic fractures. {A, B, CR}

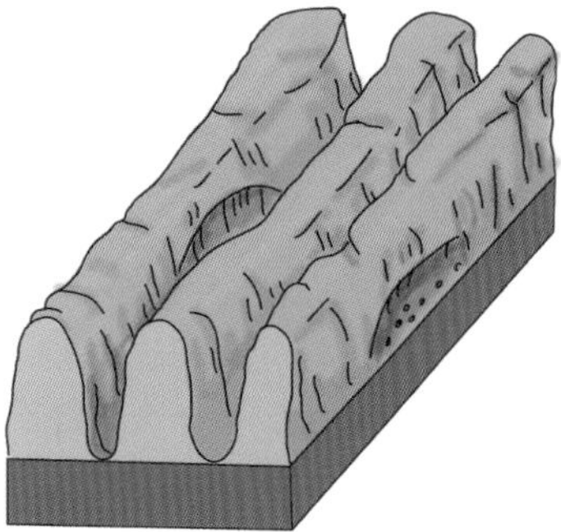

Fluvial
Depositional environments and processes involving rivers and streams and their geologic products. Cf: depositional environments. {All}

Fold
A planar structure such as rock strata that has been deformed, bent, or curved. Often a product of compressional stress produced by plate tectonic forces. Fold types include, anticlines, monoclines, synclines, etc. Cf: anticline, monocline, plate tectonic forces, strata, syncline. {CR, A, G}

Formation
A body of rock with distinctive features such as mineralogic composition, texture, fossils, or organic content, which is laterally extensive enough to be mappable. The root word of a formation name is often the characteristic lithology of the rock unit (e.g. Navajo Sandstone, Mancos Shale, Kaibab Limestone, etc.). Geologists do not use the term "formation" to refer to a geomorphic feature or landform. Cf: lithology. {All}

Fracture
A plane of breakage in a rock caused by brittle deformation. Fractures are produced when a stress, or force, is placed on a rock body. Cf: fault, joint, systematic fractures. {All}

Freeze-Thaw Cycles
Periodic (often daily) to intermittent temperature variation in which the temperature alternates across water's freezing point. Frequent freeze-thaw cycles greatly contribute to ice wedging, a form of physical weathering. Cf: ice wedging, physical weathering. {B}

Frost Wedging
See: Ice Wedging

Giant Weathering Pits
Large circular excavations due to differential erosion of injection domes. Erosional agents for these pits may include water, wind, and biological agents. Prominent giant weathering pits can be viewed within the Jurassic Entrada Sandstone in the Cookie Jar Butte area of Padre Bay, Glen Canyon National Recreation Area. Cf: differential erosion, injection domes. {G}

Graben
An elongate block of bedrock that has been lowered relative to the surrounding rocks by faults, creating a topographic basin. The adjacent block that sits higher than the graben is referred to as a "horst". Cf: bedrock, horst. {Can}

Gravel
Sedimentary particles greater than 2 mm (0.083 inches) and generally considered smaller than 75 mm (3 inches) in diameter. {All}

Groundwater
Water below Earth's surface. It generally exists in and is transported through the pore spaces of the rocks it resides in. Cf: pore space. {All}

Gypsum Sinkhole
A deep circular hole formed by groundwater dissolution of a gypsum diapir. Prominent gypsum sinkholes are found in Cathedral Valley of Capitol Reef National Park. Cf: diapir, groundwater. {CR}

Headward Erosion
The process of stream lengthening and downcutting up the regional slope toward the drainage divide. Cf: downcut, drainage divide. {Can, All)

Hoodoos
Rock columns, chimneys, or pinnacles possessing bulging knobs and narrow recesses. They are common in Bryce Canyon National Park, and were formed by differential erosion along fractures of the strata. Cf: differential erosion, fracture, pinnacles, strata. {B, A}

Horst
An elongate block of bedrock that has been raised relative to the surrounding rocks by faults, creating a topographic high. The adjacent block that sits lower than the horst is referred to as a "graben". Cf: bedrock, graben. {Can}

Ice Wedging

A type of physical weathering in which water percolates into fractures or pores before freezing. As the water turns to ice, it expands about 9%, which exerts a force of about 110 kg/cm2. This force is approximately equivalent to dropping a large sledgehammer from a height of 3 meters. These forces are enough to fracture the rock. Ice wedging is accelerated by frequent freeze-thaw cycles. Cf: fracture, freeze-thaw cycles, physical weathering, pore space. {B, All}

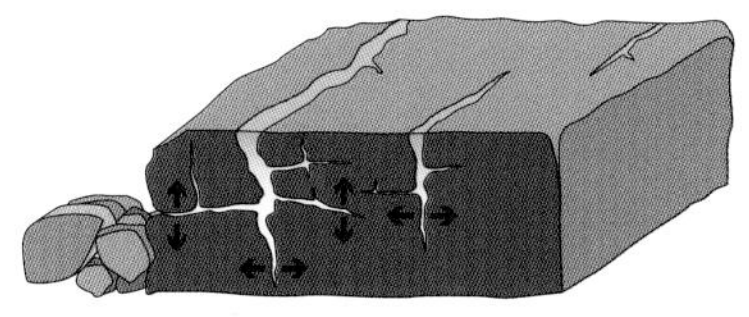

Igneous Rock

Rock formed by the cooling and solidification of magma or lava and composed primarily of minerals and glass. Igneous rocks can be "intrusive", wherein they become lithified before reaching the surface of the Earth (e.g. granite), or "extrusive", wherein the magma reaches the Earth's surface (e.g. volcano or lava flow). Cf: basalt, laccolith, lava, lithification, magma. {CR, A, B}

Impermeable

A rock unit (or some other feature within a given rock unit) that stops fluids such as water, oil, or gas from flowing through the rock. {All}

Injection Domes

A soft sediment deformational feature similar to a diapir but created from saturated unlithified sediment that is similar to the relatively undisturbed horizontal sedimentary beds that it moves through and crosscuts. Prominent injection domes, composed of mud, silt, and sand that eventually turned to stone, can be viewed in the Cookie Jar Butte area of Padre Bay in the Glen Canyon National Recreation Area. Cf: collapse and fill features, lithification, soft sediment deformation. {G}

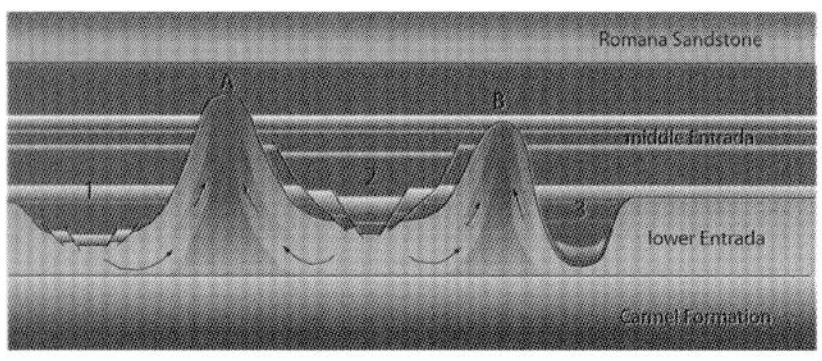

Iron

A metallic element with the symbol Fe. Iron is common in rocks and is responsible for much of the color variation. When oxidized (Fe_2O_3), iron gives the rock a red-orange-brown color. Under low oxygen conditions (FeO) iron remains black, green, or gray in color. {All}

Joint

A planar parting, crack, or fracture in a rock body wherein there is no displacement across the parting. Cf: fault, fracture. {All}

Laccolith

A large mushroom-shaped intrusive igneous rock body formed by the intrusion of magma into sedimentary rocks hundreds to thousands of feet below Earth's surface. Erosion of the surrounding sedimentary rocks leaves the laccolith exposed at the surface. The La Sal, Henry, Navajo, and Abajo Mountains of southeastern Utah are all laccoliths. Cf: igneous rock, magma. {A, CR, G}

Landslides

A general term for a type of mass movement used to describe the rapid movement of rock and soil downslope under the direct influence of gravity. Usually the moved material overlies a shear or detachment surface. Cf: curved planes, fault scarp, mass movement, rockfall. {Z, G, All}

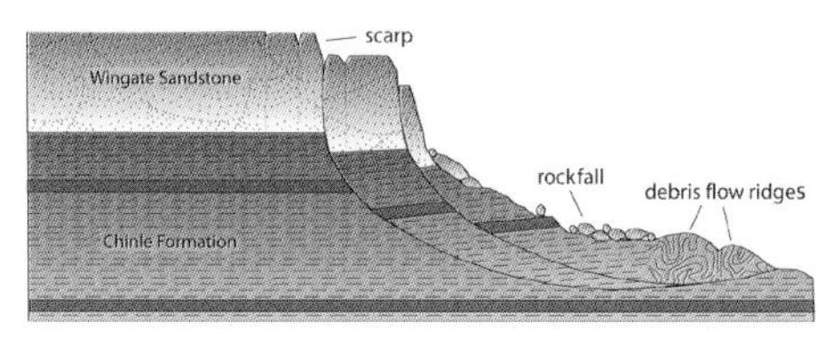

Laramide Orogeny

A mountain building event during the Late Cretaceous to Paleogene Periods (~85-35 million years ago) that created portions of the Colorado Plateau and Rocky Mountains. The event was caused by compressional plate tectonic forces wherein oceanic plates west of North America slid into and under the North American continental plate. Cf: plate tectonic forces. {CR, B, G}

Lava
Magma that flows onto the surface of Earth. Cf: basalt, igneous rocks, magma. {CR, B}

Limestone
A sedimentary carbonate rock composed mainly of the mineral calcite ($CaCO_3$). Limestones often contain fossils and commonly reflect lake deposits or low-latitude depositional environments such as the tropical and subtropical reefs of today. Cf: calcite, carbonate rock, depositional environment. {B, Can, CR}

Lithification
Refers to the process by which sediments are turned to stone. This process occurs primarily after the sediments have been buried. While buried, the sediments undergo compaction and cementation. Cf: cement, sediment. {All}

Lithology
Rock type (e.g. basalt, granite, limestone, sandstone, etc.).

Magma
Molten rock, which forms igneous rocks when it solidifies. Cf: basalt, igneous rock, lava. {B, CR}

Mass Movement
A broad term referring to the gravitational transfer of Earth material downslope (e.g. landslides, mudslides, rockfalls, avalanches, etc.). Also called mass transport. Cf: landslides, rockfall. {All}

Meander
A broad, looping bend in a river or stream. Meanders are usually created in relatively low gradient areas along the river's equilibrium profile. Cf: abandoned meander, entrenched meander, equilibrium profile. {G, Can, CR, Z}

Meteorite Crater
A near circular depression on the surface of the Earth that was caused by the violent impact of a meteorite. {Can}

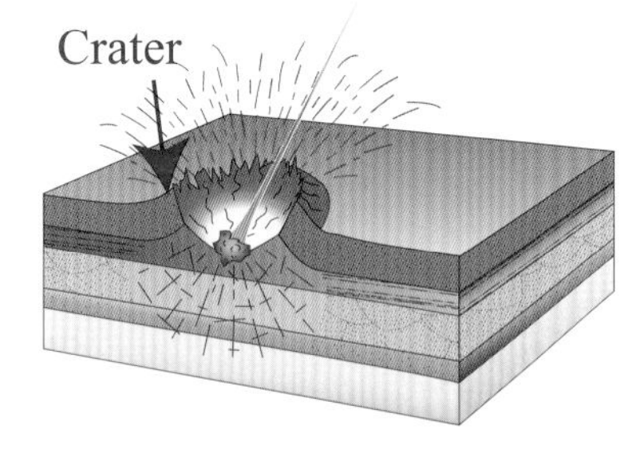

Monocline
A fold in Earth's crust, usually tectonically-induced by compressional stresses, in which relatively horizontal strata are abruptly kinked creating a steeply inclined flank. Cf: anticline, fold, plate tectonic forces, syncline. {CR, G}

Monolith
A large upstanding mass of rock (rock tower) such as those found in Cathedral Valley in Capitol Reef National Park. Monoliths are often erosional remnants of a former land surface. Cf: erosion. {CR}

Natural Bridges

Arch-like features formed by rivers or streams tunneling through underlying weaker rock at a bedding plane. This process strands the overlying bedrock into a suspended span of rock. They are often confused with arches, which form by other processes. Rainbow Bridge of the Glen Canyon National Recreation Area is a classic example of a natural bridge. Cf: arches, bedding plane. {G, B, Can, CR}

Physical Weathering

The process of physically breaking rock into smaller fragments without a change in chemical composition. Also called mechanical weathering. Cf: chemical weathering, weathering. {All}

Pinnacles

Columns of rock commonly formed by erosion along systematic fracture sets. Differential erosion of a pinnacle results in the creation of hoodoos. Cf: erosion, differential erosion, hoodoos, systematic fractures. {B, A, CR}

Plate Tectonic Forces

Massive forces caused by the movements of Earth's crustal plates. These forces deform rocks by inducing fractures, faults, and folds. It is these same forces that separate continents, create mountains and volcanoes, and continue to shape Earth's surface. Cf: anticline, fault, fold, fracture, monocline, syncline. {All}

Pore Space

Spaces within a rock that are unoccupied by solid material. This can include the space between grains, fractures, voids created by dissolution, etc. It is in and through these pores that fluids such as groundwater, oil, and natural gas are stored and transported. Pore space can be destroyed by compaction or extensive precipitation of cements. Cf: cement, groundwater. {All}

Preferential Orientation

Refers to any set of geologic features that share the same orientation in space. This could refer to canyons, fractures, pebbles, etc. {A, B, Z}

Quartz

A mineral composed of silica (SiO2). It is very resistant to erosion. It is the mineral that makes up the majority of the constituent sand grains of the many sandstones on the Colorado Plateau. Cf: erosion, mineral, sandstone, silica. {All}

Regolith

Loose debris and/or soil that overlies the bedrock. Regolith is created by physical and chemical weathering. Cf: bedrock, physical weathering, chemical weathering. {All}

Regression

Said of the sea wherein there is an increase in the area of land that is subaerially exposed as sea-level falls. Cf: transgression. {B, All}

River Gravels

Rounded gravels deposited in rivers far from their original source (tens to hundreds of miles). Increased transport distance causes greater rounding of clasts. Gravels that are deposited close to their source are more angular. Cf: alluvial gravels. {G, All}

Rock
A solid Earth material composed of minerals, rock fragments, cement or glass. Cf: cement, minerals, rock fragments. {All}

Rockfall
The most rapid type of mass movement, in which rocks of various sizes (sometimes very large) are loosened from a cliff face and fall to the base of the cliff. Cf: landslides, mass movement. {All}

Rock Fragments
Pieces of bedrock in a variety of sizes. Sandstones are composed of sand-sized (0.0625 mm – 2 mm in diameter) rock fragments and minerals such as quartz that are cemented together. Cf: bedrock, cement, quartz, sandstone. {All}

Salt-Cored Anticline
Folds of sedimentary rocks that are superimposed overtop of previously emplaced salt walls. Salt Valley and Cache Valley of Arches National Park were formed on the collapsed crests of salt-cored anticlines. Cf: fold, salt walls, anticline, and collapsed salt-cored anticlines. {A}

Salt Dome
The upward movement of a salt diapir that can produce a three dimensional dome-shaped structure in the overlying sedimentary rocks. Cf: diapir. {Can, CR}

Salt Wall
Elongate accumulations of salt in the subsurface that flow upwards through fractures or faults into the overlying strata. In southeastern Utah, some of these subsurface features are up to 10,000 feet (3050 m) high, three miles (5 km) wide, and 70 miles (110 km) long. Arches National Parks results, in part, because of the emplacement of salt walls. Cf: salt-cored anticlines. {A}

Sand Dune
A hill of sand ranging in size from tens of centimeters to tens of meters high that can be formed by flowing wind or water. Dunes are commonly asymmetrical and longer on the windward side. Sand grains are pushed up the longer windward (stoss) side of the dune before avalanching down the steeper (lee) face of the dune thereby creating cross bedding. Cf: cross beds. {Z, A, Can, CR, G}

Sandstone
A sedimentary rock made up of sand grains (0.0625 mm – 2 mm in diameter) that were cemented together usually in the subsurface. Sand grains are composed of minerals and rock fragments. The cement is most commonly silica or calcite. Sandstones reflect a variety of depositional environments including (but not limited to): beaches, rivers, deltas, deserts, etc. Cf: calcite, delta, lithification, minerals, rock fragments, sedimentary rock, silica. {All}

Scarp
See: fault scarp

Sediment
Solid particles or clasts (including minerals and rock fragments) that originate from weathering of rocks. Sediment comes in a variety of sizes such as clay, mud, silt, sand, gravel, cobbles, boulders, etc. Sediment implies transport and deposition by water, wind, ice, gravity, etc. Cf: minerals, rock fragments, sedimentary rock, weathering. {All}

Sedimentary Rock
Rock formed by the accumulation and lithification of sediment. Cf: lithification, rock, sediment. {All}

Shale

A sedimentary rock made up of very fine-grained particles such as clay and mud that are less than 0.0625 mm in diameter. Shale implies that the rock has some natural parting, called fissility, when struck by a hammer. Cf: sedimentary rock. {B, CR, Can, G}

Silica

Silicon dioxide (SiO2). Quartz is the most common mineral form. Silica and quartz are a common cement for sandstones. Cf: cement, sandstone. {All}

Sill

Igneous rocks that are injected parallel to sedimentary strata typically along bedding planes. Cf: bedding plane, dike, igneous rock, laccolith, strata. {CR}

Slopes

An inclined land surface of erosion (e.g. hillside). A slope has less dip than a ledge or cliff and usually can be negotiated by humans without ropes or climbing aids. Slopes are commonly formed by the erosion of relatively less resistant rock and/or loose sediment. Contrast with cliffs, which are formed by the erosion of stronger, more resistant rock units. Cf: cliffs, dip, erosion. {All}

Soft Sediment Deformation

Wrinkles, folds, and other warping of sediments after they have been deposited but before they have been completely lithified. Agents of deformation include gravity, fluid flow, liquefaction, fluid escape, etc. Cf: lithification. {G, A, Can, CR)

Strata

Layers of sedimentary rock (plural of stratum). Cf: rock, sediment. {All}

Stratification

The formation, accumulation, and deposition of materials into layers. A succession of sedimentary layers or beds. {All}

Strike

The compass bearing of a horizontal line that does not gain or lose elevation on an inclined surface such as a dipping bedding plane. Strike is perpendicular to dip direction. Cf: bedding plane, dip. {All}

Strike Valley

A valley eroded parallel to the strike of the strata. In an area with tilted strata, erosion will erode some rock units faster than others. These weaker units erode into valleys between the layers of more resistant rock. Cf: differential erosion, strata, strike. {CR, G}

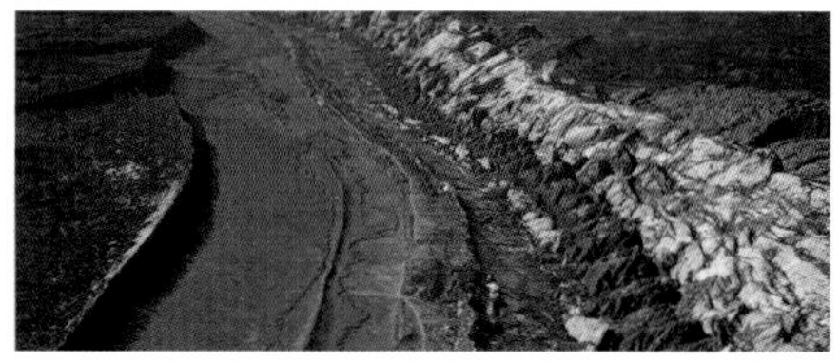

Subaerial Exposure

Sediments or rocks at Earth's surface and in contact with the atmosphere. Cf: rock, sediment. {All}

Syncline

A fold, usually induced by plate tectonic forces, in which the limbs dip toward the axis (hinge or center-line) of the fold. In folded sedimentary rocks that have been eroded flat, the younger rocks are toward the fold axis or hinge. Cf: anticline, fold, monocline, plate tectonic forces, sedimentary rock. {CR, B, G}

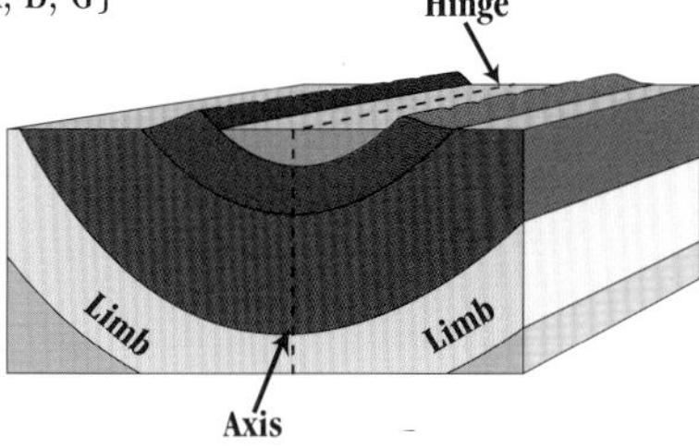

Systematic Fractures

A set of fractures that share similar orientation and spacing. It is implied that systematic fractures were created by the same tectonic event. Cf: fracture, joint. {A, B, Can, Z}

Tectonic Forces

See: plate tectonic forces

Transgression

Said of the sea wherein there is a decrease in the area of land that is subaerially exposed as sea-level rises and covers the land. Cf: regression, subaerial exposure. {B, All}

Uplift

Plate tectonic forces often cause rocks to be raised relative to their previous position and adjacent land surfaces. The uplifted rocks may then experience increased rates of erosion. Tectonic uplift has produced the Colorado Plateau. Cf: erosion, plate tectonic forces. {CR, All}

Waterpockets

Erosional surface depressions (pockets) that form on relatively flat lying rocks. Created commonly on sandstone surfaces as rainwater dissolves the cement between sand grains, allowing them to be blown away. Cf: cement, chemical weathering, erosion. {CR, All}

Weathering

The process of physically breaking, or chemically altering rocks by exposure at Earth's surface. This is not to be confused with erosion (although they work together), as erosion involves the removal and transportation of the weathered material. Cf: chemical weathering, erosion, physical weathering. {All}

ACKNOWLEDGEMENTS:

It has oft been said that we all walk in the footprints of those who have gone before us. This is true of the Geological Sciences as well. Students begin to understand geology when they realize that videos of plate tectonic re-enactments did not develop in the studio one afternoon but instead represent the culmination of decades and even centuries of painstaking field work by sedimentologists, stratigraphers, structural geologists, geophysicists, mappers, and many other thoughtful scientists. Today's geoscientists owe a debt of gratitude to those who have gone before including those break-through researchers and dedicated instructors who have given us our present understanding. Our hope is that as present-day geoscientists, we may also add to the body of knowledge and contribute to our understanding of Earth.

We express our sincere gratitude for those instructors, mentors, and colleagues who have given of their knowledge and direction. We have learned much about the geology of Utah from our colleagues at the Utah Geological Survey. In writing this book we have benefited greatly from the Utah Geological Association's 2000 Centennial Volume, Publication 28, on the Geology of Utah's Parks and Monuments. The authors' include: Hellmut H. Doelling - Arches; George H. Davis and Gayle L. Pollock – Bryce Canyon; Donald L. Barrs - Canyonlands; Thomas H. Morris, Vicky Wood Manning, and Scott M. Ritter - Capitol Reef; Robert F. Biek, Grant C. Willis, Michael D. Hylland, and Hellmut H. Doelling – Zion; and Paul B. Anderson, Thomas C. Chidsey, Jr., Douglas A. Sprinkel, and Grant C. Willis – Glen Canyon. Robert Eves' book "*Water, Rock, & Time: The Geologic Story of Zion National Park*" was also very helpful. Rick Stinchfield provided a thorough edit of the book.

We also express our gratitude for those who worked with us to develop the "GEOLOGY UNFOLDED" brochures from which we drew much to create this book. We have greatly benefited from the support of the Department of Geological Sciences at Brigham Young University. Kirsten Thompson was particularly helpful given her talent in graphic design. Laura McLevish provided stylistic comments. We recognize the Executive Directors of the Natural History Associations of the various parks for their support and efforts. These include: Gayle Pollock – Bryce Canyon; Cindy Hardgrave – Canyonlands; Shirley Torgerson – Capitol Reef; Kirk Robinson – Glen Canyon; and Lyman Hafen – Zion. Diane Allen and Murray Shoemaker of the National Park Service at Arches were equally helpful.

Finally, we express our gratitude for parents and loved ones for the opportunities afforded us and for your continued support.

NOTES:

NOTES:

ABOUT THE AUTHORS:

Thomas H. Morris

Tom's interest in geology was kindled during his "baseball days" as an undergraduate at Brigham Young University. Coming from the Midwest, he was impressed with the variety of landforms that the Rocky Mountain states displayed during his travels with the baseball team. As a junior, he took an introductory geology course and was hooked for a lifetime. Tom received his M.S. and Ph D. from the University of Wisconsin-Madison during the early eighties and then went south to Louisiana for three and one half years of working in the oil and gas industry. He made his way back to BYU as a clastic sedimentologist and stratigrapher in 1990. Tom has spent more than twenty years studying and teaching about the sedimentary rocks and national parks in Utah. His passion is showing both undergraduate and graduate students the wonders of sedimentary geology ... and Utah is one of the best places on Earth to do that! Along with his twenty plus graduate students and numerous undergraduate students, he has published more than twenty five articles on the rocks of the Colorado Plateau. His enthusiasm for solving puzzles within the Colorado Plateau seems to be ever escalating.

Scott M. Ritter

Scott M. Ritter is a professor of geology at his undergraduate alma mater, Brigham Young University. Dr. Ritter received his Ph.D. in 1986 from the University of Wisconsin-Madison specializing in applied paleontology. He taught at Oklahoma State University from 1986 until 1991, at which time he returned to Utah. His research interests include 1) the study of microfossils and their use in refining and correlating Late Paleozoic strata on a regional and global scale and 2) the geology of valley glaciers. He has conducted research in North America, Kazakhstan, Russia, Switzerland, and Africa. His favorite geological playground, however, is the State of Utah. He has traveled most of the paved roads on his Harley and many of the less accessible byways in his Jeep. A good pair of hiking boots has done the rest. From the outcrops of the Basin and Range to the slot canyons of the Colorado Plateau, he enjoys discovering the science behind Utah's scenery and sharing that with his students.

Dallin P. Laycock

As a native of southern Alberta, Dallin Laycock spent much of his childhood amongst the glacial lakes and snowy slopes of the Canadian Rockies. This led to a curiosity about the origins of the various rocks and landforms. Dallin subsequently pursued a B.S. in geology at Brigham Young University. As un undergraduate, Dallin received eleven different scholarships and awards. At BYU, he worked extensively with Dr. Morris and Dr. Ritter, and helped them complete the "Geology Unfolded" pamphlet series (which eventually led to the creation of this book). Working on the pamphlets helped Dallin cultivate his passion for sedimentary geology and develop an appreciation for the landscapes of the Colorado Plateau. Dallin received his Ph.D. from the University of Calgary. As a graduate student, he received the Queen Elizabeth Graduate Award, as well as awards from the International Scholarship Foundation and the National Science and Engineering Research Council of Canada.